Evolving Ethics:
The New Science of
Good and Evil

Steven Mascaro
Kevin B. Korb
Ann E. Nicholson
Owen Woodberry

Copyright © Steven Mascaro, Kevin B. Korb,
Ann E. Nicholson and Owen Woodberry, 2010

The moral rights of the authors have been asserted
No part of this publication may be reproduced in any form
without permission, except for the quotation of brief passages
in criticism and discussion.

Published in the UK by Imprint Academic
PO Box 200, Exeter EX5 5YX, UK

Published in the USA by Imprint Academic
Philosophy Documentation Center
PO Box 7147, Charlottesville, VA 22906-7147, USA

ISBN 9 781845 402068

A CIP catalogue record for this book is available from the
British Library and US Library of Congress

*To Charles Darwin, Alan Turing,
and their progeny*

About the Authors

Steven Mascaro, Ph.D., earned his doctorate in Computer Science at Monash University (2009) with the thesis *Abortion, Rape and Suicide*. He is a private consultant working with Bayesian network technology and web applications and continuing evolutionary and ethical simulation research as time permits.

Kevin B. Korb, Ph.D., earned his doctorate in the philosophy of science at Indiana University (1992) working on the philosophical foundations for the automation of Bayesian reasoning. Since then he has lectured at Monash University in Computer Science, combining his interests in philosophy of science and artificial intelligence in work on understanding and automating inductive inference, the use of MML in learning causal theories, artificial evolution of cognitive and social behavior and modeling Bayesian and human reasoning in the automation of argumentation.

Ann E. Nicholson, D.Phil., did her undergraduate computer science studies at the University of Melbourne and her doctorate in the robotics research group at Oxford University (1992). She spent two years at Brown University as a post-doctoral research fellow before taking up a lecturing position at Monash University in Computer Science. In addition to her interest in ALife simulations for investigating evolutionary ethics, her research spans many areas of Artificial Intelligence, including probabilistic reasoning, Bayesian networks, planning, user modeling and knowledge engineering.

Owen Woodberry is completing his Ph.D. at Monash University in Computer Science, exploring the use of evolutionary simulation to shed light on evolution theory, focusing on the units of evolutionary selection and the evolution of aging. He also has interests in Artificial Intelligence, Knowledge Engineering Bayesian Networks, Environmental Science and Teaching. In addition to his academic duties, he works as a consultant for the company Bayesian Intelligence, which specializes in Bayesian Networks.

Contents

1 A Science of Ethics   **1**
- 1.1 Ethics 3
- 1.2 Evolution 6
  - 1.2.1 The Received View 6
  - 1.2.2 The Gene's View 8
- 1.3 Simulation 11
  - 1.3.1 Artificial Life Simulation 12
- 1.4 Evolving Ethical Behavior 15
  - 1.4.1 The Iterated Prisoner's Dilemma 16
- 1.5 Experimental Philosophy 18
  - 1.5.1 Experimental Simulation 20
  - 1.5.2 Experimental Ethics 22
- 1.6 Conclusion 22

2 Ethics and Evolutionary Psychology   **25**
- 2.1 Varieties of Ethical Theory 25
  - 2.1.1 Virtue Ethics 26
  - 2.1.2 Deontological Ethics 26
  - 2.1.3 Consequentialism 27
- 2.2 The Value of Ethical Simulation 31
- 2.3 Evolutionary Psychology 33
  - 2.3.1 Tenets of Evolutionary Psychology 33
  - 2.3.2 Sociobiology 37
  - 2.3.3 The Debate over Evolutionary Psychology 38
- 2.4 Evolutionary Ethics 41
- 2.5 Against and for Utilitarianism 42
  - 2.5.1 Utilitarian Caveats 52
  - 2.5.2 Metaethics as Unresolvable 53

3 Simulation as Experimentation — 55
- 3.1 The Scope and Limits of Computer Simulation — 55
  - 3.1.1 What Computers Can't Do — 56
- 3.2 What Is Simulation? — 57
  - 3.2.1 A Definition of Simulation — 57
  - 3.2.2 Dynamic versus Static — 59
  - 3.2.3 Artificial Life Simulations — 60
  - 3.2.4 Another Definition of Simulation — 63
- 3.3 Homomorphic Simulation — 64
  - 3.3.1 Testing for Homomorphism (Validation) — 65
- 3.4 Simulations as Experiments — 68
  - 3.4.1 A Comparison with Real Experiments — 68
  - 3.4.2 The Epistemology of Simulation — 70
  - 3.4.3 Experiments as Simulations — 71
  - 3.4.4 Special Epistemology — 73
- 3.5 Conclusion — 75

4 Evolutionary Artificial Life — 77
- 4.1 Simulated Evolution — 79
  - 4.1.1 Genetic Algorithms — 79
  - 4.1.2 Simulating Evolution — 80
  - 4.1.3 Evolving Psychology — 81
- 4.2 Individuals, Agents and their Societies — 82
- 4.3 Some Ancestral Simulations — 83
- 4.4 A Simulation Environment — 86
  - 4.4.1 The Simulation World — 87
  - 4.4.2 Time — 87
  - 4.4.3 Food — 87
- 4.5 Agents — 88
  - 4.5.1 Birth, Age and Death — 88
  - 4.5.2 Health and Utility — 88
  - 4.5.3 Behavior — 89
- 4.6 Evolution and the Agent Genotypes — 91
  - 4.6.1 Production Rules — 92
  - 4.6.2 Decision Tree — 93
  - 4.6.3 Mutation and Meta-mutation — 94
- 4.7 Statistics — 95
  - 4.7.1 Demographics — 95
  - 4.7.2 Action Rates — 95

		4.7.3 Total Utility	95
	4.8	Conclusion	95
		4.8.1 Utilities in Agent-based Modeling	96
		4.8.2 Simulating Ethics	97

5 Experiments in Evolution — 99
- 5.1 Levels of Selection — 99
- 5.2 The Evolution of Aging — 108
  - 5.2.1 Comparing Alternative Aging Hypotheses — 112
  - 5.2.2 Species Selection — 127
  - 5.2.3 Summary — 133
- 5.3 Suicide as an Evolutionarily Stable Strategy — 135
  - 5.3.1 Simulation Design — 137
  - 5.3.2 Basic Demographics and Orientation — 140
  - 5.3.3 Experiment: The Evolutionary Stability of Suicide — 140
  - 5.3.4 Possible Causes of the Evolutionary Stability of Suicide — 143
  - 5.3.5 Alternative Explanation: Mutation Accumulation — 144
  - 5.3.6 Summary — 145
- 5.4 The Evolution of Parental Investment — 145
  - 5.4.1 Simulation design — 148
  - 5.4.2 Prior Investment Hypothesis — 149
  - 5.4.3 Desertion Hypothesis — 152
  - 5.4.4 Paternal Uncertainty Hypothesis — 155
  - 5.4.5 Association Hypothesis — 157
  - 5.4.6 Chance Dimorphism Hypothesis — 159
  - 5.4.7 Summary — 161
- 5.5 The Evolution of Utility — 162
  - 5.5.1 Utility and Fitness — 162
  - 5.5.2 Design of an Experiment — 169
- 5.6 Conclusion — 170

6 Experiments in Ethics — 171
- 6.1 Introduction — 171
- 6.2 Cooperation — 172
  - 6.2.1 Cultural Evolution and the Stag Hunt — 173
- 6.3 Altruism — 175
  - 6.3.1 Food Sharing — 177
  - 6.3.2 Altruistic Suicide — 178
- 6.4 Rape and Sexually Dimorphic Behavior — 181

		6.4.1	Theories of Rape in Evolutionary Psychology . . .	181
		6.4.2	The Controversy over Evolutionary Accounts of Rape	182
		6.4.3	The Unethical Nature of Rape	184
		6.4.4	Simulation Design	186
		6.4.5	Basic Demographics and Orientation	191
		6.4.6	Experiment: The Evolution of Rape and Dimorphic Behavior	191
		6.4.7	The Evolutionary Stability of Rape	194
		6.4.8	The Emergence of Sexual Dimorphism	195
		6.4.9	The Genetic Causes of Rape and Sexually Dimorphic Behavior	200
		6.4.10	Experiment with the Ethical Consequences of Rape	202
		6.4.11	Summary	205
	6.5	Abortion		206
		6.5.1	Abortion and Evolution	206
		6.5.2	The Ethics of Abortion	208
		6.5.3	Simulation Design	210
		6.5.4	Basic Demographics and Orientation	215
		6.5.5	Experiment: Varying After-birth Investment	217
		6.5.6	Experiment: Varying Gestational Investment . . .	219
		6.5.7	The Evolutionary Stability of Abortion	220
		6.5.8	The Genetic Causes of Abortion	221
		6.5.9	Introducing Fixed Genomes	224
		6.5.10	Experiment: Exploring the Ethics of Abortion . . .	228
		6.5.11	Summary	232
	6.6	Conclusion		233
7	The Future			**235**
References				**237**
Glossary				**259**

List of Figures

1.1	Charles Darwin.	2
1.2	Coefficients of relatedness.	10
2.1	Aristotle.	26
2.2	Jeremy Bentham's head.	28
3.1	A Glider.	62
3.2	A Glider Gun.	62
3.3	Testing for homomorphism.	65
3.4	The simulation onion.	67
3.5	The verification, validation and confirmation triangle.	71
4.1	Crossover of fixed-length production rules.	93
4.2	Crossover in decision trees.	94
5.1	States and transitions of the early group selection models.	101
5.2	Within- and between-group selection	106
5.3	The child's vulnerability bit string	117
5.4	The group relatedness resulting from varied genetic expiry age and vulnerability mutation rate.	118
5.5	The global population density plotted over time	119
5.6	Evolved genetic expiry age and frequency of death due to old age.	121
5.7	Mapping of the causal pathway between genetic expiry age and group benefits.	122
5.8	The evolution of aging rates.	125
5.9	Antagonistic pleiotropy and mutation accumulation.	126
5.10	Example interaction between two host mate signatures.	130
5.11	Example interaction between the host vulnerability signature and the parasite infection and virulence signatures.	131

5.12	The evolution of genetic expiry and number of species.	132
5.13	Number of species and evolved expiry ages.	132
5.14	Virulence and number of species.	133
5.15	General demographics in the suicide simulations	141
5.16	Suicide rates in constant-rate and seasonal food simulations	142
5.17	Age and health distributions of suicides	144
5.18	Prior investment: Evolved PI and offspring numbers by sex	150
5.19	Desertion: Evolved PI and offspring numbers	153
5.20	Paternal uncertainty: Evolved PI and offspring numbers	156
5.21	Association: Evolved PI and offspring numbers	157
5.22	Chance dimorphism: Evolved PI	159
5.23	A generic decision-making scenario with utilities.	164
6.1	Total and average utility in simulations with and without suicide	180
6.2	General demographics in the simulations with rape	190
6.3	Evolution of rape rates over time by sex	192
6.4	Evolved female and male rape rates for each parameter set	193
6.5	Offspring numbers by sex	197
6.6	Evolved rape rates by sex for different after-mating investments	198
6.7	Rape rates in compensation and male inefficiency simulations	200
6.8	Genetic rape probabilities for sex-linked and health rules	201
6.9	Rape rates in simulations with sex-only genes	202
6.10	Utilities for simulations with varying forms of rape	203
6.11	Food in constant-rate and periodic drought simulations	214
6.12	General demographics in the abortion simulations	216
6.13	Abortion rates for varying *abi*	217
6.14	Abortion rates for varying *gi*	219
6.15	Genetic map of observations vs abortion probabilities	222
6.16	Genetic map for constant-rate food simulations	223
6.17	The fixed genome structure	223
6.18	Fixed genome structure abortion rates for varying *abi*	225
6.19	Fixed genome structure abortion rates for varying *gi*	226
6.20	Fixed genome structure genetic map	227
6.21	Total utilities for simulations for varying *abi*	229
6.22	Total utilities for simulations for varying *fdf*	230
6.23	Total utilities for simulations with and without abortion	231

List of Tables

1.1	Example prisoner's dilemma payoffs	17
4.1	Properties of IPD, *Sugarscape* and our simulations	86
5.1	Simulation Experiment Settings.	123
5.2	Simulation Parameters.	128
5.3	Parameters of the altruistic suicide simulation.	138
5.4	Variables representing parental investments in the experiments.	148
5.5	Prior investment: Female vs male action rates	151
5.6	Desertion: Female vs male actions rates	155
5.7	Paternal uncertainty: Female vs male action rates	156
5.8	Association: Female vs male action rates	158
5.9	Chance dimorphism: Dimorphism in small to large populations	160
5.10	Chance dimorphism: Female vs male action rates	161
6.1	Example stag hunt payoffs	174
6.2	Altruism via food sharing.	177
6.3	Action utilities for suicide experiments	178
6.4	Parameters of the simulations containing rape.	187
6.5	Utilities and health effects for actions in simulations containing rape	188
6.6	Evolved rates of mating, eating and rape for both sexes	194
6.7	Parameters of the abortion simulations.	212
6.8	Action utilities and health effects	213
6.9	Average evolved action probabilities for each of the conditions in the fixed decision tree, along with standard deviations.	226

Preface

The seeds of this book were planted at the turn of the millenium, when a computer science student curious about the philosophical potential of computers encountered two like-minded lecturers. With their encouragement, this student embarked on a thesis exploring the potential of evolutionary and ethical simulation. These three were joined a few short years later by another student, equally curious about what simulations could say about foundational issues in evolutionary theory. After a considerable virtual journey led to the first author's successful PhD, Mark Bedau suggested the thesis might serve as the basis for an interesting book. For this suggestion, all four authors are very grateful. This book is the culmination of that suggestion, bringing together various interrelated strands of research pursued collectively by the authors. It combines the major part of the work of two PhDs, but blends and leavens the work leading, we hope, to a nourishing result.

Early on, philosophy described all of our attempts to advance the state of human knowledge. Physics commingled with biology, medicine, religion, politics and logic as well as matters now traditionally of philosophy, such as metaphysics, ethics, epistemology and esthetics. While Plato and others drew a boundary around natural philosophy, the distinction had never been methodological. After long ages, this began to change with some natural philosophers of the Renaissance striking out into new territory, enchanted with the methods of experiment. Physics left first, followed by others such as medicine and economics and, most recently, psychology. Other fields, for which experiment seemed impractical or pointless, remained — ethics among them. Our hope here is, first, to show that the experimental method is of as much use to ethics as it is physics, but, second, and more importantly, to show that simulation can act as a bridge between the analytical tradition of philosophy and the experimental tradition of science.

A question sometimes asked of agent-based modelers is, Why not use game theory rather than simulation? The question seems motivated by a persistent belief that simulation is the inferior option — that we drag it out only for pragmatic reasons, but would happily return into the arms of game theory if at all possible. The closed-form equations that game theory produces are simpler, more convenient and more certain than what we learn from simulations, so perhaps it is true that simulation should be a last resort. But, if so, it is a "last resort" with a vastly wider domain of applicability, as demonstrated clearly by the simulations in this book. To understate the matter, all living creatures — humans, animals, plants — are heterogeneous, whether across species, within species or even with a single kin group. Game theory does not even try to capture this heterogeneity — and if it tried, the result would surely end up indistinguishable from agent-based simulation.

Simulation research is still quite young and the practical limits of computing — power, memory and software methods — will decide how well our programs model the physical systems of interest to us for many years to come. The simulations in this book certainly reflect this, trading off detail for practical performance. We simulate as we are able; but it is clear that the most important, insightful and even groundbreaking simulations are ahead of us, and not behind us.

This work is aimed at an assortment of readers: philosophers, evolutionary biologists, economists, sociologists, psychologists, computer scientists, simulationists and the generally curious. The book is written to allow readers with different backgrounds and interests to dive in where they wish. A glossary at the back (with first occurrences of entries bolded in the text) may also help with this. Most will want to begin with the discussion of Chapter 1. Anyone wanting a review of ethical systems, and especially of utilitarianism, should look to Chapter 2; common arguments against utilitarianism are given in that chapter, along with our rebuttals. Chapter 3 argues the case that there is little difference between computer simulations and experiments, either methodologically or epistemologically. Chapters 4 through 6 then present our experimental work. A brief history of Artificial Life (ALife) is drawn at the beginning of Chapter 4, followed by a description of the common design elements of our otherwise variegated simulations. Chapter 5 applies evolutionary artificial life simulation techniques to the investigation of some fundamental issues of evolutionary theory, including the levels of selection debate. Our analysis in this chapter highlights the relation between group and kin selection, usually held to be antagonistic, but which we find to be supportive. Chapter 6 turns to a small selection

PREFACE   xvii

of ethical and unethical behaviors. While investigations of cooperation and altruism are very common, as we discuss, we also explore two behaviours that have not been deeply explored via simulation: rape and abortion.

Many of the curlier details of the simulations have been glossed over for this presentation. The reader wanting more information is encouraged to explore our papers and technical reports on the topic, which are available from our website:

```
http://www.csse.monash.edu.au/evethics
```

Also available from the website is a simplified demonstration simulation in Netlogo for each of the major simulations in the book. We hope that you will be inspired to replicate, critique and expand the scope of these simulations, and, ultimately, to contribute to the growing use of simulation and computers in philosophy, science and ethics.

Acknowledgments   We thank Alan Dorin for the cover image of our home world. Some of the ideas appearing here were tested at the Center for Logic and Philosophy of Science, Tilburg University, the Institut d'Histoire et de Philosophie des Sciences et des Techniques, University of Paris, and the 4th Australian Conference on Artificial Life, Melbourne, 2009; we are grateful for those opportunities. The second author is grateful to Volker Grimm and Steven Railsback for the chance to participate in their 2008 Summer Individual-based Modeling School in Bad Schandau. Those who assisted with reviewing, discovering errors or providing other useful comments include Mark Bedau, John Bigelow, Nick Bostrom, Allie Ford, Roman Frigg, Stephan Hartmann, Erik Nyberg, Julian Reiss, Geoff Webb. We thank our editor, Anthony Freeman, for his patience.

Chapter 1

A Science of Ethics

Throughout history philosophers have studied and debated ethical questions without the help of real-world experiments. While ethical experiments could answer many important questions, most such experiments would themselves clearly be unethical. Some empirical assistance to ethical theorizing has been found in the recent past. Since Charles Darwin, many have found the story of the evolution of cooperative and social behavior so compelling that they have claimed to find justifications for ethical behavior within that evolutionary history. This is the program of **evolutionary ethics**,[1] advocated by Julian Huxley (1927), E. O. Wilson (1978a) and many others. More empirical assistance comes from **evolutionary psychology**, which attempts to apply the concepts of evolutionary biology, and the circumstances of evolution's activity, to solving problems about current social behavior. The direct application of the facts of evolution to justifying ethical norms, however, can only get anywhere by way of the **naturalistic fallacy** of inferring ought from is. If we are to learn anything about ethics from evolution, we need a less direct route.

The very first substantial application of computer simulation was John von Neumann's simulation of nuclear reactions for the design of the hydrogen bomb, using the very first computer, the ENIAC (Goldstine, 1993). Since then, every scientific discipline — from Astronomy to Zoology — has adopted computer simulation techniques to explore beyond the limits imposed by time, money, and social and ethical constraints in a new era of scientific experimentation (see, e.g., Humphreys, 2004, Racynski and Bargiela, 2007, Frigg and Reiss, 2009). We argue that these new experimen-

---

[1]Boldface for phrases or their cognates in ordinary text indicates a corresponding entry in the glossary.

Figure 1.1: Charles Darwin.

talists are on the right track — that these computer simulation experiments have the same epistemological standing as traditional physical experiments.

In addition to experimental simulation launching new approaches to the study of the traditional experimental sciences, computer simulation has overtaken sciences which have been heavily dependent upon non-experimental techniques, such as economics, epidemiology, and sociology, offering experimental options from the 1990s. These social simulationists have drawn upon research on complex systems and the evolution of social behavior (e.g., Axelrod, 1984) to find new and fruitful applications of simulation.

We extend the application of computer simulation further. In particular, we draw together all of the above ideas into a new experimental ethics. The simulation of ethical behavior allows us to examine that behavior in ways never previously available. We can systematically alter the conditions within which ethical actions occur, the available behaviors themselves, and their impact on societies. By evolving these behaviors we can establish the evolutionary scenarios which do, and do not, support their establishment — not just cooperative behaviors, but altruistic behaviors, and selfish and other unethical behaviors as well. We are thus able to map out the evolutionary limits and possibilities for ethical action. By examining the distribution of **utilities** in the population we can also consider the moral status of contentious behaviors from a **consequentialist** perspective. All of this opens up an entirely new approach to the scientific study of ethics.

The potential for philosophical studies goes beyond ethics. There is no reason our computer simulation methods cannot be applied also to, for example, social epistemology and the philosophy of scientific method, as has already been done in preliminary forms (e.g., Hegselmann and Krause, 2006 and Gooding and Addis, 2008). We look forward to these evolving. And there is no reason questions of traditional epistemology cannot be attacked similarly, looking at the evolutionary conditions for different criteria of belief and justification to make sense. Indeed, we know of no reason why there should not be an experimental assault upon many of the traditionally a priori enterprises, investigating mind and matter, meaning and language, social and individual decision-making.

While we have a wider agenda of advancing computer simulation generally, our aims specific to this book are to introduce a new experimental science — the evolutionary **social simulation** of ethics, to justify its use as an empirical method, and to describe and illustrate its advantages. Making sense of our evolutionary simulations requires us also to make sense of some basic issues about evolution, especially why evolution works. If evolution depended solely upon natural selection at the individual level, as it would on a traditional interpretation, we would be at a loss to explain our simulation results. More sophisticated interpretations of evolution are more accommodating. So, in this introduction we shall review the basic elements of our evolutionary stories: ethics; theories of evolution by natural selection; computer simulation and especially **artificial life (ALife)** simulation; and how these are put together in the ALife simulation of the evolution of ethical behavior.

1.1 Ethics

Ethics is the study of the "shoulds" or "oughts" of human behavior. The terms 'morality' and 'descriptive ethics' are used to describe the study of a group's principles of behavior (Singer, 1994). That is, **descriptive ethics** is the study of what people in fact *believe* ought to be done. We could as justifiably locate its study within anthropology as within philosophy. By contrast, we may be interested not so much in *beliefs* about what is right, but in *what is right*. That is, we may be interested in **normative ethics** or moral philosophy (or simply ethics): the systematic study of *how we ought to act*. Normative ethics is usually investigated from within a particular ethical system, such as **virtue ethics, deontological ethics** and **utilitarianism**, to name the major players (which we will discuss in the next chapter). **Metaethics** attempts to arbitrate between these systems, aiming to supply a theory of ethical study which might give us reasons for or against the various ethical systems.

The relation between descriptive and normative ethics has been, and remains, fraught. A naturalistic ethics, as described by G.E. Moore (1903), holds that good is defined by reference to natural objects — that is, it defines *ought* in terms of *is*, effectively identifying normative with descriptive ethics. However, since Hume's (1739) argument that ought-statements cannot be derived from is-statements, philosophers have tended to steer clear of naturalistic normative theories. Moore dubbed the inference of *ought* from *is* the 'naturalistic fallacy' and accused naturalistic ethics of committing it. The relation between descriptive and normative ethics, however, is not settled by any of this. If we cannot infer the one from the other, we can certainly *inform* the one by means of the other.

There are at least two grounds for an informative relation between the descriptive and the normative. First, it is an undisputed principle of ethics that we cannot be obligated to do what we are incapable of doing. And what we are capable of doing is contingent upon many things: our physical natures, our environment and our cultures and beliefs. So, an investigation of these many features descriptive of us and our surrounds is actually essential for a proper understanding of the normative. Similar considerations have given rise to much of the interest in recent decades in "naturalized" epistemology and "naturalized" philosophy of science, both of which aim to identify normative standards for acquiring knowledge (e.g., Quine, 1969, Giere, 1985). Likewise, evolutionary ethics seeks to base ethics somehow on our evolutionary heritage (e.g., Huxley, 1927), and, while evolutionary ethics has generally been accused of committing the naturalistic fallacy, a viable alternative is to use it to help normative studies determine the boundaries of evolutionarily possible ethical behavior.

The second support for a relation between the descriptive and the normative is the metaethical process of **reflective equilibrium** attributed to Nelson Goodman (1956) and John Rawls (1972), but already to be found in a nascent form in Aristotle's *Nichomachean Ethics*. Reflective equilibrium, in brief, treats a normative theory analogously to scientific theories, with the direct empirical evidence being a core set of normative judgments common to a population. The normative theory should provide potential explanations of its evidence, our intuitions, which have the usual virtues of scientific explanations, such as being unifying (consilient), generalizing, simple, and compatible with related scientific theories. The normative theory is thus required to get considered judgments of right and wrong the same way as those judgments in the population which are (nearly) universal, but is free to carve up the remaining, unclear judgments as it will ("spoils to the victor", as Lewis, 1986, p. 203, puts it). The freedom to theorize, however, is far more constrained than that, since, as with any scientific theory, it is also obliged to be consistent with the full range of accepted scientific theories today. An ethical theory that is in reflective equilibrium today must do justice to our evolutionary ancestry, our best theories of cognition and social behavior, and much else besides.[2] These kinds of considerations are responsible in part for recent interest in a kind of experimental moral philosophy, which has aimed at identifying core moral intuitions in ways more systematic than traditional philosophy has done — that is, by substituting empirical inquiry into moral judgments for armchair speculation by philosophers (e.g., Nichols and Knobe, 2008). We applaud those efforts, and we offer another way to get out of the armchair to do empirical ethics, namely by moving over to the computer.

In order for computer simulation studies to be informative about ethics, we must adopt a point of view which allows us to measure the outcomes. Utilities are the natural currency for measuring ethical outcomes. Utilities also support a very natural ethical system, namely utilitarianism, the thesis that *what action is best collectively* is *what action is best*. Utilitarianism is, in fact, the only ethical system which *allows* us to measure the outcomes of computer simulations and judge them as better or worse.[3] And so we adopt utilitarianism both on the grounds that it is unavoidable in studies like these and because it provides a plausible candidate system for achieving the kind of reflective equilibrium theory we mentioned above. Beyond that, however, utilities provide a general-purpose apparatus for measuring the impact of actions on both individuals and societies, and they allow us to investigate the evolution of utility with potential for informing us about the evolution of action and agency (Chapter 5).

---

[2] This approach is also related to Quine's *Web of Belief* (Quine and Ullian, 1978). Much more might be said about the reflective equilibrium method. Certainly some qualifications are required. For example, the universality of opinion for the "base" cases depends upon the population of interest and for the method to have any value this must at least exclude manifestly deviant people, such as the seriously mentally ill. Furthermore, the base cases, even with an agreed population and universal agreement, cannot be treated as inviolate, any more than Karl Popper's "basic sentences" in observational science were inviolate, which is already suggested by our reference to Quine. But examination of these subtleties would take us far afield. Here we are satisfied to accept *something like* what we've outlined as a promising, indeed the best available, approach to generating a normative ethical theory.

[3] According to Stephan Hartmann, this was a point made some time ago by Patrick Suppes as well, in conversation.

1.2 Evolution

Evolution theory has become a keystone theory in science, supporting much of what we understand about life from the lowest levels of molecular biology to abstractions about human behavior, such as **altruism**, love and purpose. Evolution is also fundamental to our uses of simulation to investigate questions about biology and behavior. In order to understand this usage, we first introduce the *received view* of evolution — the standard interpretation of the "neosynthesis" of Mendelian genetics and Darwinian theory promulgated in the mid-20th century by, for example, Huxley (1942), Dobzhansky (1951) and Mayr (1976) — and its close relative, the **"gene**'s eye view" of evolution, largely based on Hamilton (1964) and developed and popularized by Williams (1966) and Dawkins (1976).

1.2.1 The Received View

Evolution has three necessary ingredients, which, when combined properly in a reproductive population are also jointly sufficient — i.e., they inevitably get evolution going:

Necessitata of evolution:

1. *Heritability.* Phenotypic traits must have a tendency to be passed on to the next generation.

2. *Selection.* Some of those phenotypic traits must have *different* tendencies to be passed on to the next generation.

3. *Variation.* The traits passed on to subsequent generations must not always be perfect copies; they must vary.

The inevitability of evolution is not a logical inevitability. Evolution is a stochastic process, subject to a great deal of randomness, for example, in mutations and other genetic modifications during reproduction and also in accidents that suddenly remove individuals from the evolutionary process through death. However, evolution is a practical inevitability: out of, say, one hundred thousand simulations of evolution incorporating the above three features in well understood ways — ways we describe in this book — we would typically find zero of them showing no evolutionary process. Creationists might regard the results of computer simulation to be no more than a "proof of concept", demonstrating only that evolution in biology is *possible.* However, the manifest applicability of the three necessitata to real biology leaves no doubt as to the reality of biological evolution.

To be sure, in a full and proper account of evolution much that is implicit in our three necessary conditions would need to be enlarged upon. For an example, the variation of condition 3 must lie within some intermediate range: too little added variation can result in selective pressures turning an initially varied population into virtual clones of each other; too much variation will kill everything off. For another example, the abiotic environment in which evolution unfolds must be neither too chaotic nor too uniform. A completely unchanging and uniform environment will not provide differential selection pressures to push evolution along. A radically changing environment, on the other hand, will not allow selected traits to be adaptive in successive environments, making cumulative **adaptations** impossible. Even more radically changing environments won't allow traits to be inherited, because they won't support continued life at all. We cannot give a full accounting here of the conditions for evolution or their interpretation; instead, we refer the reader to the neosynthetic texts cited above or to the excellent introduction to the philosophy of biology, *Sex and Death*, by Sterelny and Griffiths (1999).

"Survival of the fittest" is a slogan introduced by Herbert Spencer, meant to sum up the import of Darwin's (1859) theory (Spencer, 1864). If **fitness** is understood as 'being selected for', then, as many have pointed out, the slogan degenerates into a boring tautology. However, a more plausible and useful interpretation of **fitness** is as the number of expected descendants of an organism. Expectations are not always fulfilled, so no tautology remains in Spencer's slogan. Regardless, the slogan is suspect. It has been used to emphasize a very narrow suite of ideas about fitness, ideas of strength, speed and the physical fight for survival. No doubt, during the many millions of years of evolution, especially the recent ones, fitness has often been enhanced by improvements in communication, **cooperation**, symbiotic coadaptations, immune systems and also plumage, for those of us who must dress for success. Bloody fights to the death are relatively less common.

Computing an expected value requires having a probability distribution, and, by introducing probabilities of descendants into fitness, we allow for the tendencies (and uncertainties) of selection pressures to manifest themselves. Traits which raise the probability of reproduction will have higher fitness than those which do not. *Survival* is not actually of the essence here — a praying mantis that is likely to be eaten immediately after copulation may well have high fitness — so that is another way in which Spencer's slogan was misguided. And a final, important point about fitness is that what matters is *relative* or *differential* fitness and not absolute fitness. A spider with one million expected descendants may well be extremely unfit; it will be unfit if its conspecific spiders have ten million expected descendants. In generations to come, the traits of these latter spiders will completely swamp those that are making the first spider unfit. It is only the ratios of fitness *within* a **species** that determines which traits will win out.

In the received view the carrier of heritable traits is the genotype, chromosomal DNA. Differential selection pressure at the phenotypic level indirectly puts differential pressure on the genotypic level, with the result that the gene pool alters over time, leading generally to better adapted organisms.

This much is common to modern interpretations of evolutionary biology, even though it has become clear that the received view is too narrow. (For an important example, epigenetic inheritances outside of nuclear DNA are by now well established; e.g., Jablonka and Szathmáry, 1995.) The received view also has some commitments particular to it. Of special interest here is the idea that selection is **individual selection** — that individual organisms are the **units of selection**. Every modern theory of evolution has it that selection pressures originate with the interaction of the phenotypic traits of individuals and their environments, but it doesn't necessarily follow that the best description of what is being selected for and against *is* the individual. In fact, that commitment by the received view leaves it at a considerable disadvantage to its relative, the "selfish gene" view of Dawkins and others. Individual selection cannot explain the evolution of altruism.

1.2.2 The Gene's View

Biological altruism is any act which damages the individual's fitness to the benefit of that of others. There are many examples in nature. The actions of sentinels, in howler monkeys for example, typically put them at greater risk of predation while improving the chances of their peers to escape unharmed (Wilson, 2005). A somewhat strained attempt to explain away sentinel behavior might put it down to **reciprocal altruism**, if a round-robin or random schedule for acting as sentinel is applied. And reciprocal altruism is clearly not altruism at all, but a simple exchange of services conducted over time. But there are also more extreme forms of altruism offering no possibility of reciprocity, such as matriphagy in the *Chiracanthium Japonicum* spider, with the mother spider giving her life to foster her offspring's development

(Toyama, 2001). Altruistic behavior occurs across a wide range of species and clearly has some kind of adaptive value, but what kind? It is clearly not adaptive for the *individual*, since the individual fitness is what is being sacrificed. If individual fitness exhausts the fitness story, then altruistic behaviors which reduce it must over time simply be erased from the gene pool, so altruism can only ever evolve away. For the received view, biological altruism is just an anomaly.

Hamilton's (1964) **kin selection** theory, expanding fitness to **inclusive fitness**, provides a clear and persuasive explanation of the evolution of altruism. Inclusive fitness adds together the individual fitness effects (f_i) of an **allele** (a), weighted by Wright's (1922) coefficient of relatedness to the allele's owner (r_i):

$$F(a) = \sum_i r_i f_i$$

Individual fitness requires i to range over the actor alone, when, of course, $r_i = 1$. Inclusive fitness allows i to range over a whole population, with r_i reflecting the probability of alleles being shared. Thus, inclusive fitness measures the expected number of copies of a gene within the gene pool over future generations. So, genes with positive inclusive fitness may be expected to spread and become established in a population, and so also altruistic actions, insofar as they are genetically predisposed. As J.B.S. Haldane famously quipped, "Would I lay down my life to save my brother? No, but I would to save two brothers or eight cousins" (see Figure 1.2).[4] In Chapter 5 we will illustrate the evolution of altruistic suicide through kin selection.

Richard Dawkins (1976) has taken this idea of fitness for genes (alleles) and built up an entire *Weltanschauung* out of it, the world as seen by a "selfish gene". From the gene's perspective, organisms are simply means to an end, that end being to replicate oneself. Genes are in control, and bodies are like vehicles that they drive around. Of course, it isn't really about

---

[4]If we were to take this story exactly literally, apparently we would end up being more closely related to monkeys than our siblings, since reportedly we share 98.8% of our DNA sequence with the former (Chen et al., 2006) and supposedly we share 50% with the latter! The explanation, of course, is that the coefficient of relatedness is the probability of the two alleles being direct copies of an allele of a common ancestor in the system portrayed (i.e., with "root" nodes (e.g., AUNT) counterfactually assumed to have no common ancestry with JBSH, etc.). Since we share 98.8% of our alleles with monkeys, and 99.9% with randomly selected humans, most alleles will turn out to be the same between two humans (primates) regardless of the coefficient of relatedness between them. Nevertheless, these coefficients (at least when the system is somewhat expanded) are exactly what are needed to compute the tendencies of fitter alleles to replace less fit alleles in population genetics.

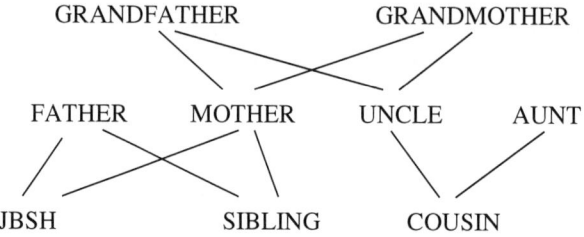

Each link weight represents 1/2 of the genes being shared
Relatedness = sum of products of link weights over all paths

$$R(\text{JBSH, SIBLING}) = (1/2)^2 + (1/2)^2 = 1/2$$
$$R(\text{JBSH, COUSIN}) = (1/2)^4 + (1/2)^4 = 1/8$$

Figure 1.2: Coefficients of relatedness.

selfish genes fighting other selfish genes for the opportunity to replicate; organisms are built by a grand assemblage of genes who all must cooperate to get the organism developed properly and so for any of them to be replicated. Other, perhaps less sensational, versions of gene selectionism are advocated by Williams (1966) and Hull (1988).[5]

Aside from its amusing new imagery, Dawkins' interpretation does not appear to be very far removed from the received view itself. It importantly locates fitness at the level of genes, which, as we've pointed out, can only be advantageous, allowing for an explanation of the evolution of altruism without losing anything, since inclusive fitness subsumes individual fitness. What selfish gene theory does not do is imply genetic determinism: although it emphasizes the role of genes in determining development, neither Dawkins nor any other advocate would claim that the genetic endowment simply overrides the developmental environment; genes clearly work within some (benign) environment to guide development. In this, Dawkins' and the received view are one. They both have been blind to non-genetic in-

---

[5]Richard Dawkins and the philosopher Daniel Dennett found this point of view so gripping that they have been driven to extend it to *ideas*, or *memes* as they would have it: memes drive humans around and make them put together compelling arguments, stories and songs so as to reproduce themselves in other people's heads (Dawkins, 1976, Dennett, 1995). It is clearly correct that ideas, beliefs, etc., as realized in human culture, possess the three necessary conditions to evolve: they are passed on from generation to generation (heritability); some are preferred to others (selection); and they are altered while being passed on (variation). Nevertheless, the added explanatory value of attributing the spread of an idea to positive selection pressure rather than an old-fashion appeal to, say, its persuasive power seems to be small, so perhaps memes about memes won't replicate much farther.

CHAPTER 1. A SCIENCE OF ETHICS   11

heritances, such as epigenetic inheritances (Jablonka and Szathmáry, 1995) and ecological inheritances (Odling-Smee et al., 2003).[6]

And both views have been hostile toward the possibility of *groups* of organisms being somehow a focus for selection pressures.[7] However, **group selection**, the differential ability of *groups of organisms*, whether tribes, communities or species, to reproduce themselves, is another potential explanation of the evolution of biological altruism. Ignoring kin selection pressures for the moment, we can agree with Maynard Smith (1976) that the individual fitness of altruists will be lower than that of selfish organisms, so that altruists will tend to die out in groups, and yet groups with altruists may well have greater expected longevities and so have greater opportunities to found new groups than entirely selfish groups. It follows that, under the right circumstances, the group selection pressure for altruists will outweigh the individual selection pressure against altruists, and so altruism will be evolutionarily stable. Maynard Smith (1976) himself, and most biologists, thought that the required circumstances are unlikely to be realized in nature. We will argue that they are mistaken and, in particular, that kin selection will often bring those very circumstances about. We will illustrate the operation of group selection, and its interaction with simultaneous kin selection, in Chapter 5.

The idea that gene selection and group selection are compatible and, indeed, simultaneously active has become respectable in recent decades (in **multilevel selection** theory; e.g., Wilson, 1997). The competition for explanatory exclusivity — between genes and groups, and between DNA and epigenetic inheritance — is an ill-conceived struggle; by dropping claims to exclusivity, all of these become compatible with each other.

1.3 Simulation

Our technique for experimental investigation here is computer simulation. So, what are computer simulations? The PC game "The Sims" is a well known example: it simulates the life and times of various characters who worry about getting jobs and cleaning toilets. Aircraft and naval piloting simulators simulate conditions involved in normal and abnormal maneuvers

---

[6]Contrary to the extremism of Fodor and Piattelli-Palmarini (2010), while the received view has ignored these sources of heritability, there is nothing incompatible with the main ideas of the received view and these additional sources of inheritance.

[7]We note that W. D. Hamilton has not shared this hostility; in fact, Hamilton (1975) gives an excellent analysis of the conditions for group selection.

of aircraft and ships. And "Second Life" simulates a large range of human and non-human activities. However, none of these are simulations in the sense we mean here: in all of them a human user plays an essential and central role, which is not to the point in simulation science.

Simulation science is about expanding our scientific knowledge, rather than entertainment or instruction. The simulations of interest to us here are those in which the *entire* simulation occurs within a computer, as a **computer process**, imitating physical or social processes in the wider world, whether chemical, astrophysical, evolutionary or ethical. These scientific simulations are commonly reported by philosophers of simulation to be nothing more than substitutes for analytic methods of solving integrations and partial differential equations, when the latter turn out to be too hard for us to do (e.g., Humphreys, 1991, Winsberg, 2001, Frigg and Reiss, 2009). In fact, however, simulations go far beyond that numerical computing role (Korb and Mascaro, 2009). Given a simulation model instantiating some theory, we can, of course, determine what the theory predicts for the future, by setting the simulation parameters to values representing the present and running it. This substitutes for deductions of predictions when deductions are hard. But we can also do many other things. We can be, and often are, surprised by what our simulations turn up: no one can envision the full range of consequences of a theory for any situation, and so no one can make predictions for all of them. In other words, beyond predicting the future, we can *explore* the future, insofar as our guiding theory is the right one. And, whether or not our theory is correct, we can explore the theory-space by altering the structure and dynamics of the simulation and observing how things would then unfold. We can also perform sensitivity analysis to determine how responsive effect variables are to initial conditions or to dynamic parameters. Numerical computation is such a narrow part of the life of a simulationist one must wonder how much actual experience these philosophers have had with computer simulation! In addition to all of those considerations, artificial life simulation in particular doesn't even begin to fit the numerical computation model, because it doesn't even begin with equations to be solved.

1.3.1 Artificial Life Simulation

Artificial life (ALife) is the imitation of living processes using computer simulation (or other technology). The term 'Artificial Life', due to Chris Langton, is just two decades old, but the idea is much older. Arguably, the first serious ALife research began in 1911, when Leduc experimented

CHAPTER 1. A SCIENCE OF ETHICS 13

with colloidal solutions to emulate metabolic functions, mitosis, and other activities associated with life (Keller, 2003). Later, von Neumann (1951), inspired by Stanislaw Ulam, invented **cellular automata** and designed the first (somewhat complicated) self-replicating cellular automaton. Self-replicating cellular automata were later made famous — and much simpler — using Conway's "Game of Life" (Berlekamp et al., 1982).

In addition to individuals, many fields have also contributed to ALife. Cybernetics passed on the study of self-organization and complex systems (Wiener, 1948). Evolutionary **algorithms** were developed as search and optimization techniques from the 1950s and 60s (Selfridge, 1959, Holland, 1975, Fogel et al., 1966, Schwefel, 1981), leading to a highly active community of evolutionary ALife researchers, including us. Finally, artificial intelligence contributed (and continues to contribute) many systems of interest in ALife, such as **Bayesian** and neural networks, **decision trees**, planning systems and more.

While dubbing the field 'Artificial Life', Langton (1989) asserted that whereas biology focuses on life as it is, ALife aims to increase our understanding of life both as it is and as it could be. That definition leaves the domain of ALife not so much open-ended as unclear. No one knows what life could or could not be. But actually, ALife is not so much a *domain* of study as it is a *methodology*, a set of techniques. Indeed, most of the researchers actively using these techniques don't call themselves ALife researchers at all: in ecology they call themselves **Individual-Based Modelers (IBMers)** (Grimm and Railsback, 2005) and in the social sciences they call themselves **Agent-Based Modelers (ABMers)** (Epstein and Axtell, 1996). Quite likely, this coyness has to do with disciplinary obligations: adopting the term "ALife" would make a social scientist or ecologist sound too much like a computer scientist. Regardless, IBMers and ABMers are all applying ALife simulation to investigate problems within their various sciences. Despite the wide adoption of ALife methods, the full range of potential applications has only begun to be explored; our exploration of ethics strikes out in a new direction.

Rather than beginning with equations to analyse, ALife researchers typically begin with complex behaviors to explain. For example, how do trout make the tradeoff between occupying rich feeding grounds and avoiding predation risk (Railsback et al., 1999)? Or, what is the relation between average household size and the spread of influenza in southern California (Stroud et al., 2007)? Or, again, why do the different sexes most commonly invest different amounts of time, matter and energy in their offspring

(Chapter 5)? In order to answer these kinds of questions, ALife modelers attempt to build simulations which show these behaviors as **emergent properties** of the system. A property is *not* emergent if it is explicitly programmed into the simulation. The underlying concern here is epistemological: we cannot learn anything from our simulations if our results have been cooked. For example, suppose we are interested in investigating the **Lotka-Volterra model** of predator-prey populations, wanting to uncover conditions under which it is, and is not, realized (Volterra, 1931). It will obviously be pointless to simply program two variables, `predator-numbers` and `prey-numbers`, and assign them the values prescribed by the Lotka-Volterra equations. The only thing we could possibly learn is whether we know how to write such a **computer program**. For an ALife model, what we shall do instead is design a program which represents some geographical region, some prey animals and some predator animals, food for the prey, and some attack and defense options for the animals. The level of detail with which we model the geography, flora and fauna will depend upon the problems of interest to us, and especially whether we are interested in very generic population dynamics or those specific to some species or habitat. However, variables reporting the population levels will have to be epiphenomenal: they must have no causal role in the simulation at all, but instead be restricted in use to generating summary statistics for output. Whether the Lotka-Volterra equations are satisfied or not will then depend upon the dynamics and conditions under which the simulation runs; it can only *emerge* as a long-range result of what we have programmed.

Many have held properties to be emergent only if they are *surprising* to the researcher. While many emergent properties certainly are surprising, this is no necessary condition. Many other emergent properties are fully expected to show up: if we have programmed our simulation right, and set initial conditions within some reasonable range of possibilities, it would simply be amazing if the Lotka-Volterra population cycles did *not* show up! Rather than grounding emergence in a subjective emotional response, a better approach is to think of properties and behaviors of the simulation as existing in layers. At the bottom layer is the program, the computer instructions we have explicitly coded. At one or more levels above that are regular properties and relations between states of the running process. They are *above* the program at least in the sense of being **supervenient** upon the program: the program realizes these properties, however, any number of very different programs could do the same (hence, **multiple realizability** is characteristic of supervenience). This account is modeled on supervenience theory in the philosophy of mind, which avoids the troubles of **reductionism** — the attempt to find bridging laws *identifying* mind and brain — by pointing out that mental properties are multiply realizable (Kim, 1993).

This yields *bottom-up computer simulation*, or "BUCS" for short (Goldspink, 2002). Epstein and Axtell (1996) suggest that BUCS is particularly valuable because it forces the researcher to look for simple explanations (in the form of simulations) for complex systems. Aside from the above epistemological concerns, another motivation is the perceived failure of top-down methods in artificial intelligence to produce a general intelligence. The hope has been that BUCS, in contrast, would be able to recreate the wide variety of living processes that exist around us, and to create new ones that do not. Whether it has succeeded is debatable (cf. Bedau, 2006); however, it has been an unquestionably fruitful approach to research.

BUCS weds magnificently with evolution: the detailed outcomes of evolution are notoriously unpredictable, indeed emergent; those which appear and reappear regularly in our evolutionary simulations become candidates for interesting emergent properties. We shall consider further how to judge the status of emergent properties, and more generally the epistemological status of simulations, in Chapter 3.

1.4 Evolving Ethical Behavior

In putting evolution and ALife together we make possible the evolution of emergent, complex ethical behavior. Of course, this is nothing new, since Evolution has already done the same Herself. The advantage of repeating the evolution of ethical behavior is that we can study the process, rather than just be its outcomes.

The study of behavior from an evolutionary perspective has been developing rapidly, most prominently in evolutionary psychology. Evolutionary psychology is based on three main principles: that behavior has an evolutionary explanation; that behavioral traits arose in an **evolutionary environment of adaptation (EEA)**, which often differs from present day environments; and that minds are modular. Modular minds can be separated into components that have evolved partially independently of each other, much as a genome that can be separated into genes (Fodor, 1983).

Evolutionary psychologists have been investigating a wide range of ethical behaviors, including those that we investigate here. The ethical studies of evolutionary psychologists, however, are limited to observational measurements and theoretical reasoning. Thus, for example, de Catanzaro

(1995) analyses suicide notes for signs that the act is aimed at benefiting kin, in accord with kin selection theory. Thornhill and Palmer (2000) propose evolutionary hypotheses (the adaptive and **by-product** hypotheses) for the existence of rape and assess their hypotheses based on the available victim data, which is notorious for its unreliability. And Lycett and Dunbar (1999) look at abortion data for single and married women of various ages, concluding that evolution may have shaped women's decision-making about abortion. All of these investigations proceed within the observational and deductive realm; evolutionary ALife simulation permits the addition of experimentation.

1.4.1 The Iterated Prisoner's Dilemma

Perhaps the most popular and commonly simulated model for ethical behavior is the Iterated **Prisoner's Dilemma** (IPD), from the game-theoretic work of Dresher and Flood (Dresher, 1961). The IPD simulations were the first simulations of significant interest to biology, and particularly to evolutionary psychology.

In the basic (non-iterated) Prisoner's Dilemma two prisoners are separated and each given two options: to rat out the other prisoner (defect), or to keep mum (cooperate). We are to assume that the prisoners have no means of communication and do not have pre-existing, binding commitments to each other. The payoffs to each prisoner of choosing cooperation or defection will depend on what the other prisoner chooses. In the original story these payoffs are reductions in prison terms (so '2' represents 2 years off), but more generally these should represent the **agent**'s utilities, so that, for example, any psychological or social disvalue in ratting someone out is already accounted for. Table 1.1 shows one possible matrix of payoffs for two prisoners, Alice and Bob.[8] The first value in each cell's pair indicates the payoff to Alice given the choices taken by both prisoners, and the second value indicates the payoff to Bob. We can see that Alice should defect if Bob cooperates, because it would pay her more (2 instead of 1). Furthermore, we can see that Alice should still defect if Bob instead chooses to defect, because it again would pay her more (0 instead of -1). Thus, defection is the **dominant strategy** for Alice, and by a symmetrical argument for Bob as well — that is, defection is *always* preferred regardless of the

---

[8]Note that there may be quite different payoff matrices for this situation, some of which will lead to different conclusions.

CHAPTER 1. A SCIENCE OF ETHICS

other prisoner's choice.[9]

| | | Bob | |
		Cooperate	Defect
Alice	Cooperate	(1, 1)	(-1, 2)
	Defect	(2, -1)	(0, 0)

Payoffs: (Alice, Bob)

Table 1.1: The prisoner's dilemma. Rows represent Alice's choices, columns Bob's choices. Values are: (Alice's payoff, Bob's payoff).

Paradoxically, despite the domination argument, double defection yields a smaller payoff to each than double cooperation; but, since it is the dominant strategy (and they can't communicate with each other), both players will defect regardless. However, it is possible that players will choose a different strategy under the *iterated* version of the game, when they must make the same kind of choice, say, 100 times in succession. The results of prior rounds will be informative about their opponents, and so they offer a form of communication. And indeed, when Axelrod and Hamilton (1981) hosted a computer tournament in which submissions were subjected to a round-robin of IPDs with other submissions, the winning strategy was not always-defect, but tit-for-tat. The tit-for-tat strategy involves cooperating on the first turn and then reciprocating whatever choice the opponent made last, cooperating when the other player does, punishing the other player when not. It seems that in a fairly wide variety of environments (types of opponents) tit-for-tat does best in accumulating utilities.

Axelrod (1984) went on to confirm this result, evolving the tit-for-tat strategy in a **genetic algorithm**, in a process he called the "evolution of cooperation". In the right circumstances not only does tit-for-tat evolve, but it can be an **evolutionarily stable strategy** (ESS). For example, a population of suckers (always-cooperate) can be invaded, and eliminated, by defectors, but a population of tit-for-tatters cannot.

This is not yet the evolution of very interesting ethical behavior: the goal of the evolved "organisms" is not, for example, altruistic, but strictly selfish, maximizing their *own* utilities over time. The fact that tit-for-tat helps other tit-for-tatters maximize their utilities as well is an incidental by-product, so calling this behavior *cooperation* is already rather optimistic —

---

[9]This also implies that defection by both is the only **Nash equilibrium** for this game.

it is based more on the tags chosen for the actions than on the meaning of 'cooperation'! Attempts in these terms to explain genuine cooperation — where the activating goal is mutual — are much like attempts to explain altruism in terms of Trivers' (1971) concept of reciprocal altruism: they are aimed at explaining them *away*. However, we shall argue that altruism (and cooperation) are real enough, and that they can be explained in evolutionary terms. And we will back these claims with our simulations in Chapter 5.

1.5 Experimental Philosophy

Experimental philosophy has suddenly become a big business. There are experimentalist philosophers investigating, amongst many other fields, epistemology (Bishop and Trout, 2005), philosophy of language (Machery et al., 2004), philosophy of science (Stotz and Griffiths, 2004) and also ethics (Knobe and Doris, 2008).[10] Part of what has stimulated this is a long-developing revolt against intuitive analytic philosophy, in particular the idea that philosophers have privileged access to concepts and ideas through their intuitions, so that the first step in analytic philosophy — getting clear about the concepts involved in some philosophical problem — can be done in the comfort of the philosophical armchair, running little thought experiments. (Of course, the succeeding steps can be done in the armchair as well, since they are deductive — indeed, as the joke goes, without even the need for a wastepaper basket.) Dennett (1991) denounced philosophical thought experiments which pretend to offer insights about circumstances with which we are wholly unfamiliar, for example, "brains" that replace neurons with humans and neural firings with instructions on paper. The experimental revolutionists go farther, demanding that the intuitions of philosophers be dethroned entirely and replaced by the intuitions of the masses. This shouldn't be a surprising step. We might, in fact, wonder why it has taken so long to arrive. Introspective psychology, for example, which supposed that psychologists have accurate insights into the workings of their own minds, died out in the 1920s, under joint pressure from the Behaviorists and the Freudians. Philosophers, no doubt, have no greater claims to special insight than psychologists. And analytic philosophers on the whole are a very unusual class of people, self-selected, well-educated and (once they get tenure[11]) well looked after. It's hardly likely that the ideas of ordinary discourse and folk psychology, the meat and potatoes of human ideation, are going to be

---

[10] For a collection of recent experimental philosophy, see Nichols and Knobe (2008).

[11] Excepting Australia, where tenure has been denatured.

CHAPTER 1. A SCIENCE OF ETHICS 19

properly served up by the likes of them!

So, experimental philosophers have the common aim of substituting empirical evidence about human ideation for unreliable, biased philosophical intuitions about our ideas. Many also take reflective equilibrium seriously as a useful approach towards explaining whatever intuitions one finds in field work. We share these interests and aims, however, we wish the experimentalists would go a little farther than they have towards *experimentalism*. In particular, most "experimentalists" are fully satisfied with opinion surveys about ideas, substituting some more representative sample of intuitions for the unrepresentative sample of one armchair occupant. But a sample survey is hardly the same as a controlled experiment.

For example, Marc Hauser (2006) conducted an on-line survey confirming armchair intuitions about the morality of killing and allowing killing. Around 90% of respondents thought it permissible to divert a tram headed for five innocent victims onto a track with one innocent victim; on the other hand, 90% also thought it impermissible to make the apparently same kind of utilitarian decision when it involved shoving an innocent fat man in front of the tram, with his death stopping the tram ahead of the five innocent victims. The best explanation for such intuitions seems to involve a contrast between allowing someone to die and being an active agent in his death. This intuition, if real, poses a *prima facie* problem for utilitarians. But the very first thing we should do in response to this is to question the reality of the intuition; there is no experimental evidence supporting its reality. Sample surveys are one thing and experiments another. To put this more pointedly: people like to think well of themselves. They spend much of their time and energy building up a world view and a view of themselves, and the latter is almost invariably positive. You may think Adolf Hitler was the epitome of evil, but it is clear that Hitler himself thought no such thing. This (extremely strong) tendency is revealed in asking people to remember long-past incidents occurring within the presence of others: no two stories are alike, and they often differ in predictable ways, those ways favorable to the story teller (as in Kurosawa's film *Rashomon*). In short, asking someone to imagine what they would do in circumstances C almost invariably results in that person reporting what she or he thinks sounds best. Putting that person *into circumstances C* will often yield entirely different results. Experimental philosophy could, apparently, learn a lot from recent experimental economics and, in particular, from adopting genuinely experimental methods.[12]

---

[12]We will grant that the proposed experiment introduces additional complexity, such

By the way, this criticism is not merely an evasion by some utilitarians of an unpleasant counterexample. The methodological defects of experimental philosophy need to be addressed by all, regardless of the subdiscipline or point of view. And, we shall defend utilitarianism in any case, even against possible counterexamples of the above type, in Chapter 2.

1.5.1 Experimental Simulation

One of the major reasons why simulation methods have gained such prominence across the sciences is that they allow easy, and easily controlled, access to experimental techniques. It is far easier to manipulate variables within a simulation than within the process being simulated. Very often, there simply is no possibility of manipulating anything in the worldly process. Ignoring our simulations, then, the only kind of empirical evidence achievable is observation. So, for example, until the rise of computer simulation, empirical astronomy was almost purely an observational affair. Now, however, simulations of the origins of solar systems and the deaths of stars are commonplace.

What can be learned from observation is substantial. What can be learned from experimental intervention is far more substantial. This is quite intuitive. Observed associations between types of events commonly lead us to hypothesize a causal connection between them: smoking and lung cancer, CO_2 and global warming, asteroidal impacts and mass extinctions. In each of these cases, and many more, the observed associations, while supporting the hypothesis of causality, failed to settle the causal question — there were (or are) seemingly endless debates with skeptics. Experimental interventions, of course, could settle the causal questions. Interventions on humans smoking, the earth's atmosphere and asteroids may not be practical options, but they are options in principle. Indeed, the smoking question has been settled, with experimental interventions in test animals helping to rule out skeptical doubts, such as Fisher's (1957) suggestion of a common genetic cause for smoking and cancer. That asteroids have caused at least some of the mass extinctions has been thoroughly confirmed by a vast array of observational data, leaving no reasonable alternative explanation for

---

as questions about the moral courage of participants. However, our point is in part that survey samples are not as simple as they seem and are not clear and direct measures of the conceptual structure of the subjects. In particular, they are subject to systematic biases. Therefore, alternative measurements, and especially measurements that probe beyond the subjects' models of themselves, are going to be useful, whether or not they are also difficult to use. There are no simple answers to difficult philosophical problems.

CHAPTER 1. A SCIENCE OF ETHICS   21

the K-T boundary extinction of the dinosaurs and half of all other species (Ward, 1995). So, that's a win for observation, but it came at the cost of massive amounts of observational and theoretical research over a decade (the 1980s). The CO_2 cause of global warming is also, belatedly, coming to public consensus, well after computer simulation experiments put the issue beyond doubt for specialists. The theoretical and simulated causal connection between rises in CO_2 levels and terrestrial temperatures is far more compelling than the associational evidence alone of a past correlation between them (Randall et al., 2007).

Observations by themselves cannot reveal the difference between a correlation of effects of a common cause, e.g., genes causing both smoking and cancer, and a correlation arising from causation, e.g., smoking causing cancer. This is a problem of underdetermination, of multiple viable hypotheses remaining after the evidence comes in. Evidence from experimental interventions can eliminate far more hypotheses than can observational evidence. Indeed, in ideal circumstances, interventional evidence can uniquely identify the true hypothesis, simply eliminating the underdetermination problem (Korb and Nyberg, 2006). Experimental simulation opens up the possibility of gathering interventional evidence when otherwise only observational evidence would be available, whether because of physical, social or ethical limits upon those interventions.

So, experiments are a major part of what simulationists do. Making sense of experimentation and experimental methods, and especially how and why we can learn from them, will be an important topic in this book. This is necessitated by a frequently skeptical reaction to the possibility of learning *anything* from computer experimentation. In the various sciences in which computer simulation plays an important role (that is to say, in all sciences) the same skeptical concerns can be raised. They, however, no longer play a serious role in discussions in experimental physics, cosmology, chemistry, chemical engineering, etc. On the other hand, they do continue to be pressed for ALife simulations which attempt to inform us about the real world — ecological IBMs, social scientific ABMs and evolutionary simulations. What we will find in Chapter 3 is that there is no interesting epistemological difference between ALife simulations and the other simulation sciences: it is incoherent to accept the experimental verdicts of astrophysical simulations and reject those of ALife simulations, so long as certain preconditions are satisfied. Perhaps more surprisingly, we shall find that there is no interesting epistemological difference between simulated experiments in any of the sciences and real-world experiments in those sciences. The calls of some philosophers of simulation (e.g., Winsberg, 2003) for a radically new epistemology to fit the radically new method of simulation will be firmly rejected.

1.5.2 Experimental Ethics

What is true of simulation across the sciences is true of simulated ethics, only more so. The kinds of experiments we perform with the virtual beings of Chapters 5 and 6 are not the kinds of things readers will want to try at home, outside of their computers! To be sure, most of them would be impossible, but many of the possible ones would be abhorrent. However, by conducting genuine experiments about rape, abortion, suicide and so on we can discover much of interest. We can discover environments where such behaviors have, and do not have, adaptive value. We can learn something about how and why they are adaptive, when they are — and vice versa. And, under the hypothesis of utilitarianism, we can learn about the morality of the behaviors under different evolutionary scenarios. There are, of course, substantial limits to what we can learn from the simulations we have performed. Some practical limitations arise from the fact that our work here is largely preliminary. This is a new field, and we are, at best, a scouting party. For an important example, our simulations here are fairly generic: we have not attempted to simulate any particular species, but rather the behavior of some very large class of species. We have aimed to cast light on general questions about the evolution of altruism and selfishness, **parental investments** and neglect. Inferences about any *particular* species will have to be qualified by considerations about their particular circumstances. We hope that others will pick up where we have left off and start simulating such particular circumstances. There is only knowledge to be gained.

1.6 Conclusion

We endorse a more scientific philosophy, one that sees little or no difference between the conjectures and theorizing of biologists and those of philosophers of biology, beyond disciplinary conventions and habits. Every science has its philosophy: physics and the philosophy of physics; AI and the philosophy of AI; psychology and philosophical psychology. Applied ethics has ethics, and ethics has metaethics. And every theory has empirical consequences — or else, as Karl Popper pointed out, we have a worthless theory. So, one of our goals is to liberate philosophy, to get it out of doors, whether literally, as in experimental philosophy, or virtually, as in our experimental ethics.

This is not Scientism, the worship of all things pronounced Scientific. There is plenty that is wrong in science, just as there is in philosophy. But we think a genuine union of science and philosophy will do far more good than harm. To be sure, some of the worst philosophy has been done by scientists waxing philosophical. And some of the worst science already has been done by philosophers waxing empirical. The disciplinary divides probably cut deepest in methodological practices, and so mistakes are most likely when researchers try out new methods for new kinds of problems. But the presence or probability of error is no reason to abandon new methods; it is rather an opportunity to learn from them.

In conclusion, we hope this trek through artificial evolution and ethics will be both enlightening and entertaining. To enhance both aspects, we make available a variety of simulations illustrating our experiments at

`http://www.csse.monash.edu.au/evethics`

These are written in NetLogo, a user-friendly computer simulation language also available on the net at

`http://ccl.northwestern.edu/netlogo`

Reading the rest of this book. The next three chapters can be thought of as introducing the experiments of two chapters following them, which lie at the conceptual center of our work. Chapter 5 presents our experimental investigations of some of the key ideas in evolution theory, including levels of selection (gene, individual, group and species) and the evolution of altruism. Chapter 6 reports on the more ethically oriented experimental simulations. Preparatory to them, Chapter 4 describes the common structure and operation of most of our simulations, and so is a necessary prerequisite to understanding them. The next two chapters, on the other hand, can stand on their own. Chapter 2 introduces ethics and evolutionary psychology and defends our preferred ethical system, utilitarianism, against some misunderstandings and some counterarguments. While our experimental work collects and reports on utilitarian statistics, the defense of their use may be skipped by the uninterested. Chapter 3 defends the view that we can learn about the real world from the virtual world; those who need no persuasion, and who feel no urge for a dose of epistemology, may easily pass over it.

Chapter 2

Ethics and Evolutionary Psychology

Before looking at computer simulations, in this chapter we look at some of the ethical theories relevant to their interpretation, as well as the most prominent research program pertinent to the evolution of ethical behavior, evolutionary psychology.

2.1 Varieties of Ethical Theory

Descriptive ethics studies what people actually believe about what ought to be done as well as how they actually behave, while normative ethics aims to determine how we *ought* to act, regardless of whether or not we do act in that way (Singer, 1994). Normative ethics in the small, in reference to dealing with particular moral issues such as abortion and taxation, is applied ethics. Normative ethics, whether in particular application or otherwise, is usually pursued from within a larger ethical system for deciding how we ought to behave. Perhaps the three most important ethical systems today are virtue ethics, deontological ethics and utilitarianism. And the process of trying to decide between these larger scale systems, or more fundamentally of trying to decide *how* to decide between them, is metaethics.

To acquire a certain amount of shared vocabulary with the reader, we now briefly describe some of the more notable varieties of ethical theory.

Figure 2.1: Aristotle.

2.1.1 Virtue Ethics

Virtue ethics comes from Aristotle and has been supported in recent times by many philosophers, including Anscombe (1958), Foot (1978) and MacIntyre (1981). Virtues are human behavioral traits that are considered good, either in themselves or through their consequences. Traits such as honesty, courage, loyalty and generosity normally count as virtues. Thus, those who exemplify these virtues in their characters will count as good, rather than those who perform particular deeds or adhere to their duties. Acts and duties are not irrelevant to virtuousness, but character and virtue precede and determine their moral worth.

2.1.2 Deontological Ethics

Deontology has two closely related senses. One is that good inheres in specific acts, and the other is that good inheres in some set of duties, rules or rights. Classical examples of the latter are found in most religions, such as the Ten Commandments of Judaism, Christianity and Islam or Jesus's moral statements in his Sermon on the Mount.

A deontology inherent in specific acts is exemplified by Kant's categorical imperatives, for example, the idea that telling the truth is an unconditional good (Kant, 1909). These contrast with hypothetical imperatives, which are conditioned upon particular circumstances. Restricting ethical principles to categorical imperatives gives ethical principles the status of natural laws, holding regardless of how humans and their customs might change. Kant held that categorical imperatives make it impossible to treat humans as means to our ends (which thought has led to a common objection to utilitarianism, which we treat in §2.5).

Kant developed a single universal categorical imperative, from which all moral duties were meant to derive, namely: "Act only according to that maxim whereby you can at the same time will that it should become a universal law." In other words, act according to those principles that are **universalizable**. We appeal to universalizability whenever we ask, "What if everyone did that?" A positive example of a universal maxim is "if you make a promise, keep it," since it is possible for everyone to follow the maxim while keeping the value of promises intact. The corresponding negative example is "if you make a promise, break it," since, if everyone followed the maxim, promises would lose their value.

2.1.3 Consequentialism

Consequentialism is a kind of opposite to deontology: rather than goodness inhering in acts or duties or rules, it holds that goodness is to be found in the *consequences* of acts, duties or rules. An early example of consequentialism is that of Plato's Socrates, who in the *Republic* suggested that breaking a promise to return a borrowed sword — when the owner is enraged and likely to use it to ill effect — would be moral. This example is a direct criticism of Kant's idea of universalizability and an indirect criticism of any deontological system, suggesting that hideous consequences may always outweigh the inherent good in following a rule.

Utilitarianism is the most widely known and debated variety of consequentialism, as well as its original form. It is founded on the idea that the consequences that matter are the utilities that actions (or rules) have for those who are impacted by the actions (or rules). What utilities themselves are is not exactly clear. Inexactly, they form a measure of the subjective value of an agent's state, which are positive for states of happiness, pleasure, love, satisfaction, contentment, etc., and negative for states of discomfort, anxiety, pain, fear, etc. Utilities can also be treated as the values of a primitive (undefined) function in the theory of decision making under uncertainty, as we shall see below. Utilitarianism advocates choosing that action which affords "the greatest happiness for the greatest number" (Hutcheson, 1738);[1] that is, actions are meant to maximize expected utility

---

[1] The better known formulation of this as "the greatest good for the greatest number" is due to Bentham (1789).

Figure 2.2: Jeremy Bentham's head.

over a population. This view is sometimes confused with hedonism, or pure selfishness.

Hedonism (from the Greek word for pleasure) is commonly attributed to Epicurus. Epicurus held that the highest form of happiness combined modest pleasure with the absence of pain, and today hedonism is understood to involve maximizing pleasure and minimizing pain. We may generalize the concepts of pleasure and pain to that of utility. We can do this by quantifying them, equating pleasure with positive utility and pain with negative utility, and ascribing utilities to other things (such as satisfaction, happiness, etc.).[2] (We could alternatively call this **egoism**, following Sidgwick, 1907.) Thus, the maxim of hedonism can be put as follows: act so as to maximize your personal sum of utilities.

Given this maxim, the *value* of an act can be expressed simply as the personal sum of utilities derived from that act, and hedonism directs us to choose the act which has greatest value. If we were *certain* about the effects of an act, we could express the hedonistic value of an action so:

$$v(a) = \sum_k u_e(e_k) \qquad (2.1)$$

where a is an action, e_k is one of the (certain) effects of a, u_e is the actor's subjective utility function that maps those effects to utilities, and, finally,

---

[2]Alternatively, we might take pleasure to *include* such things as satisfaction, happiness, etc.

CHAPTER 2. ETHICS AND EVOLUTIONARY PSYCHOLOGY 29

v is the hedonist's subjective value function that maps actions to values. In order to simplify future equations, it will be useful to group all of the effects e_k of an action in a set called an outcome o of a:

$$o = \{e_k\} \tag{2.2}$$

and have a subjective utility function, u, that maps outcomes to utilities:

$$u(o) = \sum_k u_e(e_k), \text{ for all } e_k \in o \tag{2.3}$$

If we substitute this into (2.1), we have:

$$v(a) = u(o) \tag{2.4}$$

Recognizing that the actor will not be certain what outcomes are possible, we need to alter the hedonist maxim to include this uncertainty: act so as to maximize your *expected* personal sum of utilities. This normative statement of hedonism corresponds to a common definition of rationality (e.g., Parfit 1984, Russell and Norvig 2010). We can now adjust (2.4) to incorporate uncertainty:

$$v(a) = \sum_j u(o_j) p(o_j|a) \tag{2.5}$$

where each outcome o_j is a possible set of effects (i.e., a possible world) stemming from an action, and $p(o_j|a)$ is the probability of the outcome o_j given that the action a is chosen.

Utilitarianism is an ethical system that suggests we maximize the sum of expected utilities *across a population*. There are numerous specific kinds of utilitarianism, some based on differing notions of utility, others based on how best to act according to utilitarian principles in practice. For example, act and rule utilitarians dispute the role of rules or principles in guiding ethical action, with rule utilitarians advocating adherence to rules or principles which tend to produce greater total happiness. Act utilitarians advise judging actions on their individual (consequential) merits, rather than on their overall effect when repeated (universalized, for example). When a generally good rule clearly fails in a particular circumstance, however, rule utilitarians will give up their argument; but also, when attempting to compute the expected utility of an action yields an inferior expected utility to using a rule (for example, by the delay implied by the calculation), act utilitarians will opt for the action of rule-following. So, perhaps there is no consequent difference between these two varieties of consequentialism in the end. In any case, we shall adhere throughout to an act utilitarian interpretation, at least on the grounds that our simulations are not capable of deciphering natural language rules.

Act utilitarianism subsumes hedonism by iterating the selfish computation over the entire population, thereby rendering it unselfish; or, as Sidgwick put it, utilitarianism is hedonism universalized (Sidgwick, 1907):

$$v(a) = \sum_i \sum_j u_i(o_j) p(o_j|a) \qquad (2.6)$$

where i represents an agent (e.g., a person, an animal, etc.), u_i represents the utility function of the agent i (according to the decision maker!), and v is the utilitarian value function. Of course, this is not precisely universalized hedonism, for both the utility function and the subjective probability function that jointly determine the expected value of an action *belong* to someone, namely the decision maker. The utility function in (2.6) in each application to an agent i is the *decision maker's* attribution of utilities to that agent, and, of course, the decision maker may well be mistaken about what utility the agent has under the circumstances being contemplated. In short, (2.6) is the decision maker's best *guess* about how to maximize utility over the population. Note also that by advocating the maximization of *expected* utility, this account is entirely forward looking: it is only the consequences for future utilities that matter to the utilitarian judgment.

In this approach to utilitarian ethics, ethical problems and their solutions are matters of individual judgment. Each individual agent is confronted with choices and must choose an action amongst them at each decision point in its life. If a utilitarian, then the agent will do its best to understand and represent the utilities of others in its own terms (on its own utility scale), but it will never actually have direct access to anyone else's utility function and so questions about whether different actors' utility functions are commensurate simply do not arise. Utilitarian ideas *could* be applied to social decision making, as in companies or nation states, but that would lead to a very different variety of utilitarianism from that which we consider here, and any difficulties particular to such attempts will not be addressed by us. We will consider objections and rebuttals to individual utilitarian ethics in §2.5.

2.2 The Value of Ethical Simulation

Matters of normative ethics form an underlying concern for us, but the tools we employ, computer simulations, of themselves offer little in the way of the analytic insights traditional ethicists aim for. Our simulations afford us the possibility of *testing* normative ethical judgments in applied settings. But the applications involved are far from traditional applications, since the settings are 100% contrived and under our control. This last point immediately raises questions about the epistemic virtue of this kind of investigation: in ordinary experimental science, scientists may well control everything *except* one experimental (dependent) variable of interest; the dependent variable is controlled by nature, not the scientist, leaving a reasonable hope of learning something about the world. In our in silico experiments, however, we control *exactly everything*, apparently leaving us with nothing to learn! But having raised the issue, we will now ignore it; we ask the reader's indulgence in simply sharing our belief in the possibility of learning from simulations until we tackle the issue directly in Chapter 3.

Under this indulgence, our simulations allow us to negotiate the gap between descriptive and normative ethics. Our artificial agents suffer pain and hurt and enjoy gustatory and sexual pleasure, in simulation. In different scenarios, their behavior promotes or undermines their simulated pleasure, according to their choices. Since we measure the utilities associated with the states our agents go through, the results of our simulations are directly telling for utilitarian ethics: choices that promote collective utility are good and those that undermine it are bad. Thus, we have simulation tools that provide direct tests of many of the claims of utilitarians, and their antagonists, whenever our simulations arguably satisfy the contextual assumptions of those claims. Where normative claims are made on the basis of enhanced utility, it should be possible to construct simulations demonstrating the enhancement. Perhaps of more interest than that is the fact that it will also be possible to construct simulations that *fail* to demonstrate any enhanced utility. The differences between the former and the latter are likely to tell us something about the range of circumstances within which the utilitarian claim is valid. By tuning simulations to fit the ethical beliefs of a community, we may obtain a richer descriptive ethics than that which simply records those beliefs; by looking at the utilitarian implications of different ethical choices, we may obtain new insights into what is, and is not, ethical from the utilitarian perspective.

Some of the value of this kind of utility-driven simulation derives from utilitarianism; that is, some of the value of the simulations evaporates if utilitarianism is wrong. We believe utilitarianism is right, and defend it in §2.5 below. Even many who reject utilitarianism accept much of its import and, in particular, the value of promoting general welfare, e.g., Foot (1978) and Scheffler (1982). In any case, there is direct value in these simulations not dependent upon any ethical theory. Using them to test, and map, the consequences of utilitarian ideas is itself an example of that value, enabling some amount of experimental investigation of metaethical questions about utilitarianism, of interest both to adherents and critics. It is, unfortunately, not possible to use this kind of simulation to also test deontological or virtue ethics: such systems depend upon the exact semantics of the deontic principles or the virtues, respectively, and incorporating semantic understanding into artificial life simulation in any kind of sophisticated way requires a prior solution to the problem of natural language understanding in artificial intelligence — which is to say, it presupposes that we have solved the problem of producing a general artificial intelligence, which is not a likely prospect for the near term.

As an illustration of the kinds of problems that would arise with trying to simulate deontological ethics we can take the famous case of the robots of Isaac Asimov. They were manufactured to adhere to the **Three Laws of Robotics** (e.g., Asimov, 1954, without Law 0):

0. A robot may not harm humanity, or, by inaction, allow humanity to come to harm.
1. A robot may not injure a human being or, through inaction, allow a human being to come to harm.
2. A robot must obey any orders given to it by human beings.
3. A robot must protect its own existence.

where each law is conditional upon fulfilling all earlier laws in the sequence. As a candidate deontological ethics, this is pretty sad, since it describes rules promoting enslavement rather than happiness or anything uplifting. But the point here is that most of Asimov's robot stories revolve around plot twists based upon ambiguities and misinterpretations of these rules. The very large corpus of Azimov's robot stories is testimony to the difficulties involved in nailing down even seemingly clear-cut principles. And the many interpretations and reinterpretations were what led the story arc to require the introduction of Law 0, in fact.

No such semantic understanding is presupposed by simulating utilities attached to the states of agents. The only real opportunity for disagreements or misunderstandings to arise in utility-driven simulation is in the assignment of specific utilities to specific outcomes. If those assignments fail to preserve the values (or attributed values) of some system of actual agents, then the simulation may mislead us when we attempt to formulate opinions about the real system based upon the simulation. But that is just an example of the more general concern about the epistemology of simulation: what simulations can or cannot tell us about reality. We defer all those issues to Chapter 3.

2.3 Evolutionary Psychology

Evolutionary psychology, and its ancestor **sociobiology**, have played major roles in recent debates about the evolution of humanity and ethics. Here we outline these disciplines and describe some of the key concepts to whose understanding they have contributed.

2.3.1 Tenets of Evolutionary Psychology

Evolutionary psychology aims to understand the modern human mind in the light of evolution. It originated in the 1980's (Cosmides and Tooby, 1987, Dupré, 1987), though many of its principles go as far back as James (1890) and, indeed, Darwin (1880). The founding idea is that evolution produced not just what can be seen of us, our bodies, but also our psychologies, and there must therefore be evolutionary (adaptive) explanations of our behavior at some level (Tooby and Cosmides, 1992). Few dispute this idea, although there is much dispute over the *extent* to which adaptive explanations are available, both for human minds and more generally (we shall take up that debate in Chapter 5). However, most proponents of evolutionary psychology (e.g., Tooby and Cosmides, 1992) more controversially deny a set of traditional ideas nurtured in the social sciences — ideas that Tooby and Cosmides (1992) group together under the label 'the standard social science model' (or, the SSSM).

According to the standard social science model, the mind evolved to be free of evolutionary constraints several million years ago. Now, today, the mind is a general learning and reasoning organ that works by forming associations and using rough rules of logic and probability. Evolutionary psychologists, such as Pinker (2002), depict proponents as believing in a mind that is a 'blank slate' at birth, ready to be filled with knowledge (i.e., John Locke's *tabula rasa*). Psychological explanations causally flow from the outside in, with culture entirely responsible for inscribing the mind's slate — and, therefore, for our behavior.

There are many reasons to object to the SSSM. Foremost, no doubt, is the grossly implausible idea that evolution and natural selection even *can* shut down in anything like our current (or recently past) circumstances. As we noted in Chapter 1 we have lived, and still live, in a world which instantiates the preconditions of adaptive evolution. Culture, even civilization, does not have the power to turn off differential selection pressure (at least not yet!). There are also many particular problems that can be raised for the SSSM proposal. For example, human infant behavior is left inexplicable and improbable. Babies' abilities to breathe, cry, and feed are not driven by a blank slate; if they ever had been, we would not be around to debate it! Nor are the well-documented, universal abilities of young children to acquire sophisticated syntactical rules (Chomsky, 1959). In defense against these points, the boundaries of the mental could be redrawn to exclude such non-culturally conditioned abilities, but only at the risk of emptying the SSSM of all of its content.

As an alternative to the SSSM, evolutionary psychology proposes a view grounded on three principles: the modularity of mind, the environment of evolutionary adaptation (EEA) and **universality**.

The Modular Mind

Evolutionary psychology holds that the mind inherits constraints that promote or prevent particular behaviors. Specifically, Tooby and Cosmides (1992), following Fodor (1983), suggest that the mind is made of *specialized modules* or *faculties* (**mental modules**), with natural selection acting semi-independently on each.

A mind composed of specialized faculties would be far more useful than a monolithic mental faculty. For one thing, modules allow for easy fault detection and trapping, and so also for the continued operation of fault-tolerant processing, as computer engineers have long known. Additionally, specialized modules give natural selection a finer level of resolution to work on. Proposed modules include language use (Chomsky, 1959, Pinker, 1994), facial recognition (Gazzaniga, 1989), reasoning (Cosmides, 1989), mathematical concepts (Geary, 1996) and emotions (Buss et al., 1992); many of these, and others, have considerable neurological and fMRI evidence supporting their modularity. It doesn't follow from any of this, of course, that some mental abilities may not also be fairly general.

One possible example of a module may have been identified in experiments based on Wason's (1966) selection task. This task involves a logical puzzle, that asks the subject when a conditional (i.e., an 'if-then' statement) needs to be checked for correctness. Generally, the subject is given four cards corresponding to the four possible Boolean assignments for P and Q in the proposition $P \Rightarrow Q$. Each card represents one truth-value assignment, with the assignment for P written on one side, and the assignment for Q written on the other. The subject is presented with the four cards, showing P, $\neg P$, Q and $\neg Q$, on one side each, respectively, while the other sides remain hidden. The subject is then asked which cards need to be flipped to determine if they satisfy $P \Rightarrow Q$.

The correct answer is to choose only those cards which may be showing P and $\neg Q$, since $P \wedge \neg Q$ is the only assignment that can falsify $P \Rightarrow Q$. When the selection task involves abstract ideas such as letters and numbers, subjects often fail to choose $\neg Q$ or choose Q instead. Many researchers have found that a subject's performance improves markedly when the task is reframed in a less abstract, more familiar context. Cheng and Holyoak (1985) suggested that people employ pragmatic schemata geared to particular kinds of problems, rather than rules of logic; selection tasks involving permission, obligation or causation were readily dealt with. While this view can be readily integrated with the modular hypothesis, Cheng and Holyoak (1985, p. 395) themselves believed that such schemata were "induced from ordinary life experiences" due to culture, rather than inherited.

Later, Cosmides (1989) undertook another study of the Wason selection task and found that when the implication referred to a social contract — for example, when P represents 'cost paid' and Q represents 'benefit accepted' — then subjects are much more likely to choose correctly. Cosmides argues that social contract theory subsumes much of the previous work done on the Wason selection task, but the most critical difference between her work and that of others is her suggestion that cheater detection is itself an *evolved* module, not the product of a general inductive mental ability.

Many have criticized the idea of a modular mind, while Tooby and Cosmides have moderated the claims in their earliest work (Cosmides and Tooby, 2000, Over, 2003). Nevertheless, the modularity hypothesis remains a central and influential theory in evolutionary psychology. In our simulations many of the behaviors are modularly driven and separately evolvable.

The Environment of Evolutionary Adaptation (EEA)

Evolutionary psychology also asserts that the mind is primarily a product of *prehistoric* evolution, and especially of the evolution of the genus homo during the Pleistocene. Evolutionary psychologists call this time and place, in which humans were evolving their unique abilities, the environment of evolutionary adaptation (EEA).

Today, our phenotypes (both physiological and behavioral) represent those qualities that enhanced our fitness (that is, the reproductive success of our genes) in the EEA, but that may no longer do so. A common example is our taste for sweets and fats: this was adaptive in our resource-starved past, but not for many in an overweight and diabetic present. Evolutionary psychologists believe this kind of anachronistic fitness-maximization applies generally to life in the developed world. (Indeed, there is a lay version in which people try to live something like the paleolithic lifestyle inside our big cities, no doubt spread by Facebook fan groups.)

Universality

Evolutionary psychologists also claim that there is a "universal" human nature (where "universal" means allowing exceptions!). For example, if we evolved a cheater detection module, then every (or almost every) person today has that module, in better or worse shape. There are three ways in which faculties may have become universal: the faculties may have spread by being *adaptations*, or as *by-products* of another, adaptive character (i.e., by piggybacking), or they may have become universal by *chance*. If they are adaptations, then the faculties would have made their bearers (or their kin) fitter than their peers during the EEA; such faculties would have spread across the population until everyone had them. If they are by-products, then they were linked to other, mental or physical, adaptations that spread during the EEA, though the faculties themselves did not contribute to fitness. If they became universal by chance, then they were selectively neutral during the EEA, but came to be a part of our present-day phenotype through either genetic drift or a genetic isolating circumstance in our species' history (for example, if our species' size dwindled to a small number at some point).

The most common view in evolutionary psychology is that most of our mental faculties are the result of adaptation (Fodor, 1983, Tooby and Cosmides, 1992), since the other possibilities appear too unlikely. There are, however, nay-sayers; Stephen Jay Gould, for example, has regularly suggested that many of our mental faculties are non-adaptive by-products of other traits or simply exaptations.

2.3.2 Sociobiology

Sociobiology is where this began. Evolutionary psychology inherited almost all of its ideas from sociobiology, to be sure with changes. Indeed, arguably evolutionary psychology just *is* sociobiology, but with its major flaws repaired.

Sociobiology is the study of all social behavior from an evolutionary perspective. While the term 'sociobiology' had been used prior to 1975 (Segerstrale, 2000), it was Wilson (1975) who attempted to forge the new field with his monograph *Sociobiology: The new synthesis*. In that book, Wilson synthesized the new ideas that were successfully describing social behavior from an evolutionary perspective — ideas such as Hamilton's (1964) kin selection theory and Trivers's (1971) reciprocal altruism. These are still central to evolutionary psychology.

The most controversial idea in sociobiology — what Philip Kitcher (1985) calls **pop sociobiology** — is that we can explain human behavior with genes and evolution. Of course, evolutionary psychology incorporates this same idea, but in contrast, sociobiology assumes that human behavior is fit for today's world and that genes account for very much of the variation in human behavior. In the extreme, a few sociobiologists attribute so much of our behavior to genetic causes that the label 'genetic determinism' actually seems appropriate, even though critics, especially Gould and Lewontin, have misapplied the label far and wide, branding also those who are not at all extreme.

Overall, pop sociobiology has been criticized on three main grounds: political, philosophical and methodological.

Politically, some argue that pop sociobiology can be used to justify the social status quo (Sociobiology Study Group of Science for the People, 1978), based on claims that current behavior is (close to) evolutionary optimality, and throwing in the naturalistic fallacy. Another concern is the thought that innate behavior cannot be undone, so the broader the reach of innateness the greater the pessimism over opportunities for change. This is undoubtedly the weakest class of objection to sociobiology. Political guidance to scientific conclusions has a considerable history of distorting and misdirecting science, notably including the Nazi "science" of race and the Soviet Lysenkoist alternative to evolution theory. If objectors do not like how the science is used in debates over social policy, then perhaps they should find better arguments about social policy, rather than objections to the science!

Philosophically, pop sociobiology makes a commitment to reductionism in human behavior, opening itself to the charge of ignoring human free will. Wilson (1977, 1978b) defended his reductionist method on the grounds that it is has met with success in many other fields. He argued that methodological reductionism does not commit oneself to *philosophical* reductionism — that is, the belief that all that can be said about sociology can be said in the language of biology. Unfortunately, Wilson never managed to explain how methodological and philosophical reductionism differ, which seems especially neglectful given that he argued against the mind being an independent, emergent property.

Methodologically, pop sociobiology suffers many problems. Much of the early work in pop sociobiology was characterized by anecdotal stories ("just so" stories of adaptation), lax descriptions of human populations and unsubstantiated inferences on the evolution of social behaviors. As Buss (1999) describes it, pop sociobiologists satisfied themselves with the ability to present an evolutionary, adaptive story of a behavior, but gave little consideration to alternative explanations.

Evolutionary psychology addresses many of these problems. While some central claims — such as kin selection, reciprocal altruism and parental investment — are adopted by both fields, evolutionary psychology recognizes the restrictions on how they apply to humans. Further, evolutionary psychologists are not generally as fiercely reductionist as sociobiologists, while still strongly emphasizing analytical methods. And finally, evolutionary psychology's methods are stronger, characterized by obtaining results from many different disciplines over many populations and drawing on the techniques of psychology and other social sciences.

Nevertheless, evolutionary psychology has had to fight battles of its own. We describe some of its controversies next.

2.3.3 The Debate over Evolutionary Psychology

Of evolutionary psychology's three principles described earlier, the modularity of mind is the most controversial — almost creating as much controversy as that over genetic determinism and sociobiology (e.g., Gould and Lewontin, 1979, Lewontin, 2000). A major stimulus for the controversy has been the idea that modularity smacks of reductionism, reducing the mind to a set of simple mechanisms and apparently leaving little room for free will and the proper expression of our mental potential. We do not think this kind of argument, or motivation, has much merit. First, it ignores the prospects of a non-reductive, broadly supervenient understanding of mental properties, including free will. In any case, free will, and the actual expression

CHAPTER 2. ETHICS AND EVOLUTIONARY PSYCHOLOGY

of our mental abilities, can simply be taken as established facts about our mentality. Compatibilist accounts of free will, which interpret deterministic mechanisms as compatible with all that we understand about our human will, are serious contenders for providing our best understanding of free will (McKenna, 2009). So long as that is the case, an evolutionary psychology, whether reductionist or supervenient, cannot be considered refuted by the existence of free will.

Some have thought that the modularity thesis implies a one-to-one mapping between evolved modules in the brain and behavioral traits, but that is a simple mistake (for discussion of such an error by Stephen Jay Gould see Korb, 1994). Just as there are **polygenetic** traits (a trait that involves more than one gene) and **pleiotropic** genes (a gene that produces multiple phenotypic traits), there may be *multiple* modules that produce a *single* behavioral trait and *single* modules that produce *multiple* behavioral traits.

In any case the modularity thesis, and the empirical neuroscientific research localizing various mental faculties to particular neural structures supporting it, are not intrinsically reductive: they provide micro-explanations of higher-level functions, just as biochemistry provides micro-explanations of the operations of living organisms. These kinds of explanations are compatible with the higher-level sciences being reduced to the lower-level sciences, e.g., biology to chemistry or chemistry to physics, or in this case psychology to biology, but they are equally compatible with non-reductionist views. Specifically, they are compatible with the view that the higher-level science is supervenient upon the lower-level science, implying at least that the possibility exists that quite different lower-level systems may realize the very same higher-level psychology (Putnam, 1975). For example, it may turn out that intelligent aliens exist and even that they have the very same modular mentality that we do. We will examine supervenience theory in more detail in Chapter 5.

Many objectors complain that evolutionary psychology promotes racism and sexism. While evolutionary psychologists have denounced the use of the modular model to support racial arguments (suggesting instead that modules force explanations to transcend race), in practice, modularity could easily lend itself to the positing of racial differences. Also, evolutionary psychologists often do posit behavioral differences between the sexes (e.g., Trivers, 1972, Symons, 1979, Buss, 1988), making the charge of sexism easy. Nevertheless, the modular mind thesis itself does not depend on or imply the existence of sexual or racial differences, and their use in theories that do makes for no good argument for or against modularity itself.

Ultimately, the value of the modular model depends on the practical and theoretical success it achieves: whether it can explain widely accepted facts and whether it can help discover new ones. Unfortunately, problems and polemics arise when one side believes their preferred method is meeting with success, while the other side believes it is not. For example, if women are found to prefer higher status men, then evolutionary psychologists explain it adaptively and social scientists explain it culturally. The same is true if women are found to prefer men exhibiting cues of higher parental investment. In fact, the same is true with almost any prediction made by either evolutionary psychology or the social sciences. Both adaptive and cultural explanations can be given for anything that is true today: i.e., they are both capable of producing post-hoc, imaginative fairy tales ("just so" stories).

So which explanation do we prefer? For now, the choice remains unclear, which is perhaps difficult for some evolutionary psychologists to admit. Thus, if a social scientist wants to explain entrenched sex differences today as a result of chance sex differences in the past, that is not to propose the impossible or a 'reversal of causation', as Tooby and Cosmides (1992) have described it. Rather, it is a hypothesis in need of evidence.

Perhaps stronger evidence in the future will allow us to settle which kind of explanations applies in specific cases. But we will only have this evidence if we gather it. We need the efforts of evolutionary psychologists and social scientists to do this properly, as well as that of computer simulationists. Most explanations of social and psychological phenomena will eventually need to be integrated across the sciences, but it is not surprising that scientists begin with ideologically driven views. Such beginnings are the norm for science, as are synthetic, consensual endings. Ideologies, and their counter-ideologies, drive theoretical growth and experimental endeavor. The lively exchanges within science usually prevent any one of them from achieving hegemony before some more orderly vision, constrained by reasoned criticism from skeptics and a recalcitrant world, becomes accepted. Harm arises only from occasional hegemonies within science, and from an after-life of skepticism turned into dogmatism without science.

Evolutionary artificial life provides us the opportunity to make some progress negotiating the differences between evolutionary psychologists and their antagonists, by providing experimental settings where, at least sometimes, the correct explanation of the evolution of a character may be settled, whether it is adaptive, a by-product or accidental, and when adaptive, how it is adaptive. In consequence, simulation can provide new information about the extent to which different varieties of behavior are adaptive, such as altruism, selfishness, or parental investment in offspring. They can also contribute to wider debates over the nature of evolution, including **punctuated equilibrium** theory and Gould's anti-adaptationism, as we shall illustrate in Chapter 5. The new evidence of artificial life will certainly not resolve all debates over the correct explanation of the evolution of particular traits, let alone the wider disputes over the nature of evolution, but it may well transform the debates by retargeting them at the relations between such simulations and evolutionary processes, allowing the debates themselves to be more focused and contentful and less rhetorical and dogmatic.

2.4 Evolutionary Ethics

Evolutionary ethics began with Darwin's suggestion in *The Descent of Man* (1880) that our moral sense should be studied as a matter of natural history. Thus, evolutionary ethics aims to ground ethics in our naturalistic history, but precisely what is at issue remains unclear. If the claim central to evolutionary ethics is that our ethical beliefs and behavior have evolved and have adaptive value, then it is a purely descriptive theory and is a corollary to evolutionary psychology in general. However, evolutionary ethics has frequently been put as the claim that an ethical system which has evolved and has adaptive value is thereby demonstrated to be *good*, which is a far stronger claim, one that simply embraces the naturalistic fallacy.

As descriptive theory, a central matter for the evolution of ethics is the possibility, or probability, of the evolution of altruism, for altruism (or, more generally, cooperation) is a key feature of complex ethical behavior and so something evolutionary psychology must be able to explain in evolutionary terms in order to be successful. We've already explained the potential for kin selection to explain the evolution of altruism, and in Chapter 5 we will present simulation experiments demonstrating this potential via both kin selection and group selection.

Most ethicists assiduously avoid the naturalistic fallacy of affirming an ought in response to observing an is, but some evolutionary ethicists apparently just embrace the naturalistic fallacy. For example, Julian Huxley began with the progressivism of social Darwinism and ended with normative ethics (Huxley and Huxley, 1947, p. 137):

> Evolution as a whole... is characterized by introducing the evolving world-stuff to progressively higher levels of organization.... We

can say that this is the *most desirable* direction of evolution, and accordingly that our ethical standards must fit into its dynamic framework. In other words, it is ethically right to aim at whatever will promote the increasingly full realization of increasingly higher values.

Similarly, E. O. Wilson (1975, Chapter 27) found normative implications in evolutionary facts, such as the diversity of moral practices implying a diversity of normative ethical standards.

These kinds of views are sustainable only upon the assumption that what has evolved is right for us, i.e., they live or die with the naturalistic fallacy. We shall leave them to their fate. While we are interested in extracting as much as we can from evolutionary psychology, and descriptive evolutionary ethics in particular, we do not expect this to include very much normative content.

2.5 Against and for Utilitarianism

Here we canvass some of the more popular, or controversial, objections to utilitarianism and give our replies. The complaints about utilitarianism are heard widely and loudly. But our adoption of utilitarianism here is not merely tactical, based on its utility in analysing simulations, but also because utilitarianism does the best job of making sense of our intuitive assessments of moral problems. It also is very widely misunderstood, as evidenced by many of the complaints; examining some of those misunderstandings is our task here.

There is more to life than happiness. And so what? There is more to utility than happiness, as well; or, happiness, as noted repeatedly in the history of utilitarianism, is itself much more than hedonistic pleasure.

In response to this kind of point, Matthew Ostrow is reported to have argued, in a Wittgensteinian attempt to dissolve the problem, that this kind of inflation of the concept of happiness reduces argument in favor of utilitarianism to a tautology (Wikipedia, 2010): that what gives happiness is what is desired, and what is desired is what gives happiness. The conclusion is that utilitarianism finds no justification in the philosophical analysis of these concepts.

However, that conclusion, and the dissolution, hurts only those who still believe in the strict program of analytic philosophy. Our more "naturalized"

CHAPTER 2. ETHICS AND EVOLUTIONARY PSYCHOLOGY 43

approach of reflective equilibrium is untouched by it. And utilitarianism itself certainly reduces to no tautology. Its central assertion is that ethical value derives from supporting the collective ability to maximize the satisfaction of our desires. The existence, indeed primacy, of alternatives to utilitarianism throughout human history is perfectly plain, whereas there are no alternatives to tautologies.

Negative utilitarianism. Karl Popper (1953) suggested that utilitarianism got right is aimed at minimizing harm, rather than maximizing pleasure. However, this can lead to quite different conclusions than an ordinary interpretation of utilitarianism, for example that genocide (executed "humanely" and not painfully) is one of the great acts of generosity available to us, for it reduces all possibility of future harm to nothing!

But this alternative is based upon a simple-minded interpretation of utilitarianism as "the greatest good for the greatest number". Rather than take that rubric *literally* — as its own author did *not* intend — we should take it as a demand for ethics to be sensitive to the directly perceived values and disvalues of the sensible subject to its state of being. Combining positive and negative utilities and maximizing the result strictly subsumes *both* positive and negative utilitarianism, and it eliminates such spurious conundrums as the alleged value of mass murder.

Average versus total utilities. An alternative to the standard model of utilitarianism, seeking to maximize total expected utility, is to maximize average utility. This allegedly avoids special difficulties for utilitarianism, such as Derek Parfit's (1984, Chapter 17) "Repugnant Conclusion", that large numbers of happy people (or happy goats, or simply contented oysters) are preferable to a small number of *really* happy people. But this conclusion is not obviously repugnant. In any case, simply breeding large numbers of animals (or humans) is highly unlikely to maximize long-term expected utility, both because of the environmental and political risks of large populations (a fact we seem only willing to acknowledge under duress!) and because of the opportunity cost of forgoing very substantial lifestyle improvements for smaller populations. (And whether we should be seeking the happiness of oysters or people, we shall address below.)

The average utility view, on the other hand, has utterly idiotic consequences, such as the putative moral value of killing off people with below average (but still positive) expected utilities during their future lives. Performed iteratively, this "improvement" would presumably lead to moral solipsism!

Utilities can't be measured. We know what it's like for another human to be hit on a finger with a hammer. We also know what it's like to enjoy a well-prepared meal. In general, we can analogically ascribe utilities to those similar to us, experiencing events similar to what we experience, but how much utility should we ascribe to a horse chewing hay? Advances in neuroscience promise to answer such questions in considerable detail... some day. In the meantime, the brute fact that we recognize pain, and degrees of pain, in the treatment of horses (witness the introduction of laws regulating the treatment of animals) shows clearly that we in fact ascribe utilities to such animals all the time. There is uncertainty about their mental states, exacerbated by their limited options in communicating with us, but the uncertainty is very far from amounting to an uncertainty about there actually *being* mental states for horses, excepting the most extreme anthropocentrists (or solipsists) amongst us. Further, the preferences of horses can be measured almost as readily as those of humans.

Frank Ramsey (1931) first presented the now standard technique for eliciting utilities from agents. We first identify two states (objects, prizes — anything within the domain of the utility function) for which the agent shows differing preferences, as can be determined behaviorally. Ideally these will be fairly divergent preferences (otherwise the procedure below requires the addition of some modest complications). The preferred choice is assigned utility 1 and the worse choice 0, simply to fix a scale. Now, for any new choice whose preference lies between these two a precise utility is readily assigned by finding that probability p such that $p \times u(\text{first choice}) + (1 - p) \times u(\text{worst choice}) = p = u(\text{new choice})$. That is, by finding the probability p that renders the agent indifferent between the new choice and this lottery, we have fixed the utility of the new choice on our utility scale. From this basic beginning has grown within the broader field of decision analysis a rich practice and theory of preferences, originally axiomatized by von Neumann and Morgenstern (1944), who also proved that utilities elicited via Ramsey's method produce a utility function over an interval measurement scale.

The only complication with applying this technique to horses is one of communication. No doubt some choices will be too difficult to put to horses in a meaningful way ("how do you feel when you've proved a difficult theorem in number theory?"), but perhaps these choices don't really matter for them. Communicating the probabilities of success and failure with lotteries requires persistence in presenting sample frequencies to horses. But horses, and many other animals, are capable of responding to sample frequencies in sensible ways. By developing some history of interaction with horses and observing their preferences, we will inevitably learn of correlations between behaviors tied directly to their expressed preferences and other behaviors, allowing analogical inferences similar to those we apply to other humans. In short, there is no difficulty in principle in extending utilities to other species, whether or not there are difficulties in practice in overcoming human prejudices about them.

The population is unknown. This is true. We can do utilitarian calculations for assumed populations, but we can't be sure these are the *right* populations. There are two primary sources of uncertainty here: (1) Are some among us hidden? Are we just unaware of some agents who will be impacted by our decisions? (2) What are the boundaries of sentience? Have we included all and only sentients who will be impacted by our decisions? The first category might include, for example, aliens. Even if there are no UFOs among us, our (future?) decisions may well ultimately impact upon those from neighboring solar systems. The second issue refers back to the last problem of measurement; we shall say somewhat more about it below.

Infinite sums. Nick Bostrom (2009) argues that in an infinite universe, total expected utility is infinite, so all actions are equally defensible on utilitarian grounds, having no possible effect on the total — in other words, under utilitarianism moral nihilism reigns! For this argument to work, the universe must be *relevantly* infinite; e.g., it must be infinite not just in extent, but in the number (or duration) of worlds occupied by sentient agents.

There are a number of likely objections to Bostrom's argument. For one thing, it shares some features with Blaise Pascal's infamous argument for belief in God. "Pascal's wager" proposes to analyse that belief as a decision problem: we decide either to believe in God or to disbelieve. There are two relevant states, according to Pascal, namely either that God exists or that he does not. If he exists and you choose to believe in him, then your reward is infinite utility (in heaven); if either he doesn't exist or in any case if you disbelieve in him, then you receive a finite (positive or negative) reward. Given *any* positive probability in the existence of God, the expected utility of belief is positive infinity and this is then the only rational choice. There have been many counterarguments offered, but a perfectly sufficient one is to point out that the "state" in which (the traditional Christian) God doesn't exist is massively ambiguous. There are many ways in which that God might fail to exist, and, in particular, for each way in which (a version of) that God might exist and provide heavenly rewards there is an equal and opposite (jealous) God who would provide infinitely negative punishment for belief in that Christian God. By partitioning the non-existence of the Christian God into such other possibilities, the expected utility of each choice, belief or disbelief, becomes indeterminate, and Pascal's argument fails.

Bostrom's argument relies on an infinite series of utilities experienced by agents, with a preponderance of positive over negative utilities, or, at any rate, a series whose sum diverges, whether into positive or negative infinity. But there is surely no compelling analytic argument in favor of this possibility; rather, Bostrom simply points out that there is no compelling analytic argument *against* it, and so we are rationally obliged to confer it "some" positive probability. As soon as we confer *any* finite positive probability to this kind of outcome, the expected utility of any single action becomes swamped by the infinite expected utility contributed by that future of experienced utilities independent of our action. But just as with Pascal's wager, for every possible series leading to infinite positive utility we can posit another possible series leading to infinite negative utility. Perhaps if we align all these possible series just right, their expected values, positive and negative, will meet and vaporize each other! A more likely conclusion is that the problem is massively underdetermined, leaving us with no moral to draw.

An even more telling consideration is that utilitarianism requires the computation of *expected* utilities consequent upon an action. The expectation can only be based upon the subjective probability of the decision maker. And, as the causal chain lengthens along which those consequences are contemplated, the probability of any specific outcome diminishes — that is, consequences that are nearby in space and time are generally far easier to anticipate than consequences that are remote. The exceedingly far-flung and tenuous consequences of agents in other solar systems, galaxies or even other universes rationally receive exponentially decreasing amounts of credence. It is more than plausible that even if those consequences are infinite in number, their weighted sums are not. There is an effective horizon of expectation which implies we are each operating within a finite world of foreseeable consequences. The consequentialism of utilitarianism requires us to focus on the finite bounds to the observable ripples induced by our actions, rather than the possibly infinite ocean into which they are tossed.

CHAPTER 2. ETHICS AND EVOLUTIONARY PSYCHOLOGY 47

The ends don't justify the means. The idea here seems to be derivative from Kant, who claimed that his categorical imperative makes it impossible to treat humans as means to ends, because such treatment is not universalizable. But act utilitarianism does not aim at finding rules, whether universal or conditional. If the net good of treating humans in a specific case as a means to an end is greater than that of the alternatives, then there is no reason at all to wax Kantian: the ends *do* justify the means.

The worry of the objectors in particular cases may well be the slippery-slope argument. For example, consider euthanasia. Some who object to euthanasia in a specific case of someone suffering serious harm during a terminal illness will agree that considering the act of euthanasia in isolation that act might be good. Say, if the world were about to end, ending the sufferer's pain a few minutes earlier than that might be a good thing. But they may be worried that in the *real* world one act of euthanasia will lead to others in similar cases, and these may lead to yet others in less similar cases, until finally we have euthanasia on demand.

The thing to notice about this sort of argument is that it in fact has exactly nothing to do with utilitarianism. If the argument is correct and good, it implies that the long-term expected utility of the immediate act of euthanasia incorporates a good deal of consequential negative utility that perhaps has been neglected by the actor/decision maker. In that case, utilitarianism advises the decision maker to take that all into account. If the argument is incorrect, then its conclusion can be dismissed. In either case, the real issue is not that ends cannot justify the means, but that incorrectly identified and weighed ends will lead us to the wrong means, which is not some radically new discovery that challenges utilitarianism.

Whether slippery-slope arguments work sometimes, always or never is beyond the scope of our work here.

William James' problem with the Moral Scale of utilitarianism (James, 1891).

> If the hypothesis were offered us of a world in which Messrs. Fourier's and Bellamy's and Morris's utopias should all be outdone, and millions kept permanently happy on the one simple condition that a certain lost soul on the far-off edge of things should lead a life of lonely torture, ... [would we not] immediately feel, even though an impulse arose within us to clutch at the happiness so offered, how hideous a thing would be its enjoyment when deliberately accepted as the fruit of such a

bargain?

In other words, the extreme negative utility of an individual cannot be paid for in the coin of positive utility of the remainder, no matter how many or of what magnitude, at least when the bargain is made consciously and as a single transaction. Many appear to share this intuition, as reflected in the fact that this kind of issue has arisen in fictional literature as well, for example in Dostoyevsky's *Brothers Karamozov* and Ursula Le Guin's short story "The Ones Who Walk Away from Omelas" (cf. also the "utility monsters" of Nozick, 1974).

But that response is not universal. For example, we simply deny that accepting the bargain is immoral. If there is a better bargain to be struck, then by all means strike it. Furthermore, there is significant incentive on offer actively to find or construct an alternative. But if there is none to be had, we cannot see what moral charge can then be made against those taking the original bargain.

To be sure, there may be *further* issues behind the scenes that the moral outrage of objectors is founded upon, such as slippery slopes leading to more and more innocents being tortured, but raising such issues changes the problem and therefore, justifiably, the response. Reflective equilibrium demands that universally agreed cases be decided according to that universal judgment, but it does not say that non universally agreed cases be decided by the majority vote, so we are happy enough even if we are here in the minority.

Peter Singer, animal rights and infanticide. Peter Singer begins his ethical theorizing with a thorough-going egalitarianism that applies to all sentient beings, quoting Sidgwick approvingly (Singer, 1976, p. 6): "The good of any one individual is of no more importance... than the good of any other." Of course, this puts some burden upon who, or what, is to count as an "individual", that is, what is the population of interest? Singer's answer in *Animal Liberation* is all sentient animals, animals which are capable of suffering or enjoyment, implying that they have interests which can be satisfied or frustrated. The result is that no sentient species is privileged relative to any other. Unsurprisingly, Singer is a leading light amongst animal liberationists and an antagonist of "speciesism". Also unsurprising is Singer's endorsement of euthanasia for some seriously handicapped people whose only prospect of future life involves on-going pain. Handicapped people have objected in open protests against Singer and his right to utter such things publicly. Properly construed, however, the debate should be about

CHAPTER 2. ETHICS AND EVOLUTIONARY PSYCHOLOGY 49

euthanasia for the suffering, rather than anything specifically to do with those who are handicapped. (And, of course, it should be a debate rather than a stifling of debate!)

Perhaps more surprising are some of Singer's other conclusions drawn in the name of utilitarianism. For example, he has said that given a forced choice between a human infant and a mature horse, he would save the life of the horse.[3] Despite his claim of egalitarianism between sentients, in this Singer reveals a clear preference, based on the fact that the mental life of the infant is weak, inchoate and largely potential, whereas the mental life of the horse is mature and vibrant. However, the utilitarian dictate to compute *expected* utility requires us to consider the potential mental life of the infant, and its probability of realizing that potential. By ignoring that, Singer has actually abandoned utilitarianism proper. In defense, Singer suggests that considering potential would lead us into a morass. For example, it is the potential for human life that leads the Catholic church to condemn the use of condoms for contraception. But the Catholic church is simply *not* taking the probabilities into account: there is no probability whatsoever, and therefore no potential in fact, for millions of sperm to lead to human lives in any one sex act. On the other hand, considering the overall potential for a couple to conceive and raise a family over the context of decades is precisely what the Catholic church refuses to do in ascribing all morality to the *individual* choice between using or refusing contraception. More generally, incorporating sensible probabilities into the ethical equation is something neither Singer nor the Catholic church are doing.

We think Singer is right that there is a difference in the mental lives of infant humans and horses, but that the *potential* mental lives are also different and the latter differences would almost always outweigh the current, temporary difference. Singer's choice is wrong. But we generalize from here: the richness and diversity of mental lives differs not just between horses and humans, but also between horses and pigs, and pigs and dogs, and dogs and squirrels, and all of these and reptiles, and all of them and insects, etc. We are in favor of "speciesism" if this means recognizing that there are mental, as well as physical, differences between species. Singer's "egalitarianism" is based upon an equality of interests; however, while all sentients experience *some* forms of pleasure and pain, they do not experience the same forms, nor are their experiences always rightly regarded as being of the same magnitude. Singer's preference for any individual over

---

[3] Details in this paragraph refer to a public lecture by Singer and a subsequent exchange with an author.

any other (e.g., horse over infant), without taking into account expectations about the future, reveals an underlying agreement with us on this point. But if the cumulative expected utilities of individuals from distinct species vary, then utilitarianism proper obliges us to discriminate between members of those species, so once again Singer diverges from the utilitarian standard.

Unlike Singer, then, we accept that humans have more "rights" than ants, in the sense that in the utilitarian equation the magnitude of utilities attributed to humans will often (but not always) be greater than that attributed to ants. This corresponds not only to how humans generally act and judge these relative values, but also to what we know about the neurological systems of these actors. This kind of speciesism conforms both to common sense and to science.

Singer and others may be worried about speciesism of a more pernicious kind, where animals are abused and mistreated, where their desires are given no weight. But whatever that view is, it is not utilitarianism, for denying the utilities of animals denies the common evolutionary heritage of all of us: it is either ignorant or evil.

The Three-Mile Island Effect. Dennett (1995, p. 498) points out that the long-term consequences of the accident at Three Mile Island include reactions against nuclear power plants and greater regulation, so we don't know even now whether the accident was a net benefit to society. Dennett doubts the value of an ethical theory which cannot make up its mind about particular cases. But the answer surely is that there is no ethical system, other than the most trivial and misguided, that can "make up its mind" about every case put to it. As good utilitarians, we do the best that we can at the time we have to choose. The psychological literature on heuristics and judgment strongly suggests that the best that we can do will often involve shortcuts (heuristics) for estimating probabilities and expectations (see, e.g., Kahneman et al., 1982, Gilovich et al., 2002, Gigerenzer and Todd, 1999). But this is hugely unsurprising: uncertainty and risk are inherent in reasoning under uncertainty. Certainty is at best a standard for mathematics, rather than science or ethics.

Putting utilitarianism forward as a normative theory, offering a standard against which judgments and actions can be compared, even while withholding practical advice in particular cases such as Three Mile Island, is *not* vacuous, contra Dennett. Given the normative standard we can *test* actions against it and *develop* practical advice accordingly. If utilitarianism were *always* lost at sea about any particular decision, then perhaps it

CHAPTER 2. ETHICS AND EVOLUTIONARY PSYCHOLOGY 51

would be useless, but Dennett's pessimism is founded on a particular, and extraordinarily difficult, case.

Perhaps Dennett thinks the Three-Mile Island case is *typical*, however, rather than extraordinary. If the claim is that all agents are always in the dark about the consequences of their actions over the long-term future, and therefore in no position to compute expected utility *ever*, then the argument simply misconstrues utilitarianism (again). Utilitarianism advises the agent to maximize *expected* value. The expectation is that of the agent: consequences the agent cannot expect, because it cannot anticipate them, are not part of the equation. Utilitarianism asks that you do the best that you can, not the best that you can't.

The Fat Man Argument. The fat man argument in §1.5 has been held by many to cut against utilitarianism, since the common preference to *not* shove the man onto the tram tracks appears to violate the utilitarian calculation that five lives are worth more than one. But there are at least a few substantial grounds to withhold that verdict. First, the commonality of the preference to save the fat man is considerably less than the universality of that preference. Many prefer to save the five under threat, rather than the fat man, even when this obliges them to take an active role. As noted elsewhere, reflective equilibrium does not require that a normative theory side with the majority opinion, so utilitarianism is not required to do so here, even while maintaining its reflective balance with universal intuitions. Thus, the experimental philosophy account of the ethics of the fat man is at most a descriptive account of a common intuition; it is not normative even on the basis of reflective equilibrium theory and so doesn't amount to a serious challenge to utilitarianism. But it may also be worthwhile assuaging the feelings of those uncomfortable with simple utilitarian obstinacy in the face of such examples. And there are multiple explanations of the common intuitive response compatible with a utilitarian treatment. For example, the slippery slope may again be a factor: people may well be thinking that if we allow active killing in this case, then active killing on weaker and then even weaker grounds may be the inevitable consequence. Although not explicitly put as a part of the fat man story, to the extent that such considerations enter into people's thinking, whether consciously or not, the story is other than as it has been depicted, so the simple utilitarian calculation is simply misleading. Another, and perhaps more important, point is that the uncertainties implicit in the story are being ignored in the simple utilitarian calculation. What is the probability that the fat man's body will suffice to stop the tram?

Indeed, what is the probability that you will successfully topple the fat man, rather than just pissing him off and receiving a pummeling? These things go unstated in the story, but hardly unthought of other than by the unthoughtful. The actor who doesn't consider them in considering the action at issue is either negligent or not a utilitarian.

2.5.1 Utilitarian Caveats

There are three aspects of the utilitarian calculation that are often ignored by critics:

- *Total* expected utility is what matters. Utilitarianism per se does not identify the population over which utility should be accumulated. If opinions differ about what population is relevant, it is clear that different judgments will ensue.

- Total *expected* utility is what matters. Ignoring long-term expectation is the source of much confusion. And especially ignoring the fact that the expectation can only be that of the decision maker is the source of confusion. Actors will know more about their own times and circumstances, as well as their own utilities, than those of other times, places and people. This can explain the apparent selfishness of some utilitarian choices: they are not necessarily selfish when the consequences of actions for others remain obscure. We are all far better masters of our own actions than those of others. Of course, this is no argument for steeping ourselves in blissful ignorance; acts of learning have both costs and benefits themselves, which include the improved understanding of the consequences of other actions.

- Total expected *utility* is what matters. As Bentham, Sidgwick, Singer and others have averred, pain and suffering, pleasure and joy — in all their manifold types and degrees — are what matter to us. Utilities of varying magnitude, both positive and negative, are needed to measure and weigh these experiences. They are measured for animals of all varieties, both explicitly, by testing choices and preferences, and implicitly, e.g., in our laws and regulations governing the treatment of animals. Utilities are measured; therefore, they can be measured.

2.5.2 Metaethics as Unresolvable

The basic question arising from ethics is: Why should we to do one thing rather than another? We have, collectively if not universally, abandoned the easy comfort of being told what to believe, the comfort of being told how the universe is by our parents, of being guided to a specific religious creed which then tells us what is right and wrong. But it is not clear that, having abandoned that, we have anything better to adopt. The approach of reflective equilibrium seems to be the best available. It seems that we can accommodate the universal moral judgments of our time and place using utilitarian theory, and further that utilitarian theory is consistent with our best science, including psychology, evolution theory, and evolutionary psychology. But we have no really compelling reason to believe that a moral system guided by our time and place is itself a system applicable to any time and place, and specifically to our own futures. In particular, if we encounter an alien civilization which makes systematically different judgments from ours, our ethical theories — and anything reflective equilibrium may have produced, so also our other normative theories — will come under serious pressure. So, it is hardly clear that utilitarianism is the final answer to anything. But it does seem the best we can do here, for now.

Chapter 3

Simulation as Experimentation

3.1 The Scope and Limits of Computer Simulation

Computers have rapidly become the primary intellectual tool deployed by humans. This is natural. One of the first things students of computer science learn is that computers are **universal:** within the range of computable functions, there is simply *nothing* that computers cannot do. Every normal programming language is, in fact, a universal Turing machine, as can be proved easily by programming a simple universal Turing machine in that language. This universality, coupled with the rapid expansion of computational power, means that computers support almost everything that occurs in developed economies. It also means that computers can be applied to nearly any intellectual task, if not as an independent source of innovation, a use waiting at least upon some considerable new developments in artificial intelligence, then as a helpmate and support. Computers can be, and have been, applied to further research in biology, chemistry, microphysics, macrophysics, economics, sociology, art, music and philosophy. Computer **simulation** has become a reliable and regular contributor to investigation in each of these fields of endeavor, and probably every science. This expanding reach of computation has led to extreme reactions, including those who see computation as essentially inferior to human inference, and these uses of computation as epistemologically suspect, as well as those who see no bounds to computer application and who call for a new epistemology to underwrite these activities.

We shall now attempt to develop some middle ground between the more extreme reactions to scientific computer simulations and especially to defend this widespread use of computer simulation across the sciences. In this we will find considering the relation between computer simulations and scientific experiments to be interesting and fruitful. Many within the new epistemology camp suggest that simulating "lies between" theorizing and experimenting (e.g., Humphreys, 1993, Winsberg, 2003, Rohrlich, 1991); if simulation is part theory and part experiment, then the old stories of how we learn about theory from experiment can hardly apply. However, we do not agree with this "in-betweenness" theory; rather, we suggest that the old stories about the growth of scientific knowledge, whether right or wrong about science before the computer, are equally right or wrong about current science. We begin by considering what computers and their simulations are.

3.1.1 What Computers Can't Do

There is general agreement that ordinary computers cannot compute non-computable functions, e.g., solving Turing's Halting Problem or computing the Busy Beaver numbers. To generate solutions to such problems computers would need to have access to infinite precision real numbers, for example, which is something no finite digital machine can manage. There is not any consensus about what this restriction really means, however. Some, such as Penrose (1999), seem to think this implies that computers are significantly inferior in potential computational ability to analog computers, such as humans. But for such a potential to be manifested, one must find an analog means of taking advantage of infinite precision real numbers, which presupposes overcoming quantum limits and pervasive low-level thermal noise. Since such limits have been operative throughout the entire evolutionary history of humanity, and since human mental capacities have certainly evolved largely for their adaptive value, it follows that human mentation as it currently exists has no more ability to break the barrier of non-computability than does the humble desktop computer. If there is potential to break through that barrier, we can hardly expect unassisted evolution to find it. We find fully satisfactory Turing's original answer to a similar complaint put to the possibility of machine intelligence — that any formal system is constrained to a proper subset of the truth by having a Gödel sentence true of it: who is to say we are any different? Computers are limited. But only a fool can fail to see the many severe limitations of humans.

If we find something that humans *can* do, then we have a *prima facie* case that computers can do it too. If we are too stupid to figure out how to get them to do it, that is a problem not attributable to them.

3.2 What Is Simulation?

So, what are simulations? The PC game "The Sims" is a simulation: it simulates the life and times of various characters who worry about getting jobs and cleaning toilets. Aircraft and naval piloting simulators simulate conditions involved in normal and abnormal maneuvers of aircraft and ships. And Second Life simulates a large range of human and non-human activities. Despite many commentators on the philosophy of simulation taking these sorts of cases seriously (e.g., Frigg and Reiss, 2009, Humphreys, 2004, Kueppers et al., 2006), in all of these simulations a human user plays an essential and central role, which is not to the point in simulation science. It is also not to the point that in ordinary language these processes are called simulations; that usage is simply emphasizing that humans are being put into other-than-real-world situations. Such simulations are not in general being used to expand our scientific knowledge, and so they do not raise the epistemological questions we wish to engage here. The simulations of interest to us here are those in which the *entire* simulation occurs within a computer, as a **computer process**. Indeed, we shall argue that the simulations of interest here are computer processes which simulate *other processes* — whether chemical, ecological, astrophysical or from whatever other scientific study.

A genuine example of a computer simulation — if also a very simple example — is Conway's Game of Life, which we will use to illustrate some simulation concepts below. If you've never run a computer simulation, we recommend you take time out to run this simple example. For a NetLogo version of the game see:

http://ccl.northwestern.edu/netlogo/models/Life

3.2.1 A Definition of Simulation

A commonly used definition of computer simulation is:

Definition 3.1 *A computer simulation is "the use of a computer to solve an equation that we cannot solve analytically."* (Frigg and Reiss, 2009)

See also, for example, Humphreys (1991), Pritsker (1979), Kueppers et al. (2006), Winsberg (2001).[1] A comment of Reddy's (1987, p. 162) might

---

[1] It is worth noting that Humphreys has retracted this view, finding the arguments of Hartmann (1996) persuasive (Humphreys, 2004, p. 108).

be confused with this kind of definition: "Simulation is a tool that is used to study the behavior of complex systems which are mathematically intractable." That, however, would be a confusion of an accidental with an essential property: we use tools where they are useful, and not where they can be used but are unhelpful.

Definition 3.1 itself, however, includes both too much and too little.[2] Whether "we" can or cannot solve an equation analytically is surely immaterial. For one thing, that would render the term absurdly relative to the individual; for example, many programs which for us would be simulations would not count as simulations for a John von Neumann. For another, as new analytic techniques become available, what once counted as a simulation may not any longer.

We do not want a concept of simulation which is relative to time, place or individual calculating ability; we want a concept which is secured by a methodological role within science. But focusing on the positive side of the definition above, things only get worse.

It's true that computer simulation began with the work of von Neumann, Metropolis and others working out ways of computing solutions to equations required for the development of the hydrogen bomb. This led to such procedures as "Metropolis sampling", Monte Carlo (MC) integration and Monte Carlo methods in general. Monte Carlo integration is a method of numerically solving an unanalysable (or difficult to analyse) integral;[3] it does so by averaging pseudo-randomly selected values of the function in question. One can think of it as throwing darts at a board where the curve is drawn and using the frequency of darts under the curve as an estimate of its area. MC integration contrasts with numerical quadrature, which sums the areas of rectangles bounding portions of the curve. Nobody talks of the latter as simulation, but it solves equations just as well as the MC approach (at least given moderately well-behaved curves and low dimensionality). However, under the definition above quadrature counts as simulation. And, if MC integration were to be counted as simulation, we can't see any reason to deny the application to numerical methods generally, since they are all about solving things with computers that we cannot solve in our heads.

However, we think it is far preferable to deny that equation solving is simulation and reserve that term for (computer) processes which mimic relevant features of a dynamic physical process under study (which is Hart-

---

[2] This is a point originally made by Hartmann (1996).

[3] Note, however, that unanalysability is not a part of anyone's *definition* of Monte Carlo methods; it's just that analysable integrals are analysed instead!

mann's definition Hartmann, 1996, p. 83; see also Zeigler, 1976, Pritsker, 1984). Racynski and Bargiela (2007) have recently stated this nicely in their first sentence: "To put it simply, computer simulation is a process of making a computer [process] behave like a cow, an airplane, a battlefield, a social system, a terrorist, [an] HIV virus, a growing tree, ... or any other thing."

3.2.2 Dynamic versus Static

Frigg and Reiss (2009) have objected to the idea that simulations are inherently dynamic, being processes that model other processes. It does not matter, according to them, that the computer process *takes* time, so long as it *represents* time: "All that matters is that the computer provides states that come with a time index.... If ... we have a computer that can calculate all the states in no time at all, surely we don't feel we lose anything."[4]

It is, of course, true that if the computer process encodes a representation of all the time steps of the target process, whether simultaneously or not, then it contains all the *information* that a simulation would carry or convey. However, it hardly follows that it *is* a simulation. For example, we might have a function state(cond,i) which returns the simulated process's i-th state, given initial conditions cond. This function contains all the information contained in the simulation; indeed, by iteration it could be used to run the simulation. However, we can equally well use it to run all sorts of processes which are *not* the simulation, for example, all the odd-numbered states or the states indexed by the Fibonacci sequence.

For a more homely example, a similar point can be made about a feature film sliced into individual frames and put in an album: the album is not the feature film. The album contains all and only the information within the movie. But a movie moves, an album does not; and the album will remain not-movie until someone splices it back together. The methodologically relevant point is that one can poke a computer process and *then* see what happens. That is at least part of the point when people note that simulations embody aspects of experimentation. But if the process is already completed, one cannot poke it. There is no experimental side to things, even if the **computer program** incorporates all the information of the original simulation. Since the information is in there, presumably there is some way of extracting the same information as one would from experimenting with

---

[4]We should like to point out that, despite our differences on some particular issues, and especially the definition of simulation, Frigg and Reiss (2009) present a parallel argument to our own, in particular advancing our shared claim that the epistemology of simulation is the epistemology of experimentation.

a simulation; but it would not be by some intervention which mimics experimentation. One might well say, along with Frigg and Reiss (2009), that since all the information is there, none of this matters. But keeping *some* connection with ordinary language and ordinary semantics is necessary, and calling a photo album a movie is plain silly.[5]

3.2.3 Artificial Life Simulations

A final, and decisive, objection to Definition 3.1 is that there have arisen very large regions of simulation research which are not plausibly described as equation solving at all, covering at least the vast bulk of agent-based modeling in artificial life and social simulation and individual-based modeling in ecology (e.g., Grimm and Railsback, 2005). The simulations described in this book are of this type. Although some equations will inevitably describe some characteristics of such simulations, it is at least *highly unusual* for the solution of equations to be the motivating factor in such investigations. The motivation is more typically the investigation of high-level properties of the system which are **emergent**. Some example motivations are:

- Demonstrating a feasible mechanism for the Baldwin effect in evolution — that is, the ability of individual learning to enhance the genetic fixation of new adaptations (Hinton and Nowlan, 1987)

- Showing that flocking behavior can result from independent decision-making throughout a flock of "boids" (Reynolds, 1987)

- Determining the minimal space requirements for beech forests to survive in isolated patches (Grimm and Railsback, 2005, §§1.2.2 & 6.8.3)

- Finding conditions supporting or undermining the main postulated mechanisms for the evolution of dimorphic parental investments in offspring (Mascaro et al., 2005)

- Investigating the effectiveness of different possible public health interventions in response to a smallpox epidemic (Eidelson and Lustick, 2004)

---

[5]For a final analogy, you might consider Hans Moravec's proposal for life-extension: downloading the information content of your brain into a disk and "waiting" for technological development to support "your" reanimation. We suggest the incredulity this idea induces in most people is simply rational. The process matters.

CHAPTER 3. SIMULATION AS EXPERIMENTATION 61

If the philosophy of simulation is not to be left behind by the science of simulation, Definition 3.1 must be abandoned. Therefore, we shall adopt Hartmann's definition, but rendering it more explicit below.

Emergence. In the simulations referred to in this list, each of the high-level properties are *emergent*. This means minimally that in none of these simulations were the properties of interest explicitly coded into the simulations and, in particular, there were no equations built into the simulations which guaranteed that the simulations would exhibit those properties. What we mean more exactly is that the emergent properties **supervene** upon lower level, base properties. Thus, the rules Reynolds directly implemented in order to get his "boids" (simulated birds) to flock were (more specific) versions of the following rules for individual boids:

1. Avoid local crowds of boids

2. Steer in the same direction of the local average direction of flight

3. Steer towards the local centroid

None of these rules describe or imply flocking behavior; however, in the broader context of Reynolds' simulation they result in flocking behavior. The specific versions of the rules above (i.e., with ambiguities about parameters, etc., removed) are the basic properties upon which the flocking behavior supervenes. Clearly, there will be a range of parameters all of which will still support the flocking behavior; that is, flocking is **multiply realizable** — implementable in a variety of specific forms. Surely also there are qualitatively different sets of rules which would likewise produce flocking behavior. So, supervenience implies a one-way dependency relationship: if the emergent properties vary, then the supervenience base must have varied; but when the supervenience base is changed, the emergent properties may not change, since they are multiply realizable.

To give some more simple examples, this time from Conway's Game of Life, Figure 3.1 shows a "glider", which is a configuration of cells in the Life cellular automaton which "glides" across the two-dimensional screen, reproducing itself in a cycle that moves to the "southeast". Figure 3.2 shows in the upper left a "glider gun" which shoots out gliders. The game of Life allows for these kinds of behavior — and much more — to emerge from random initial states assigned to a two-dimensional grid of cells, together with the rules:

62   CHAPTER 3. SIMULATION AS EXPERIMENTATION

Figure 3.1: A Glider.

1. Any "live" (on) cell with fewer than two, or more than three, live immediate neighbors dies;

2. Any live cell with two or three live neighbors lives;

3. Any dead cell with three live neighbors becomes alive.

These rules say nothing about configurations of cells or patterns that sustain themselves, let alone far more complex higher-level properties the Game may exhibit, including computational universality (Berlekamp et al., 1982).

Figure 3.2: A Glider Gun.

The evolutionary simulations which provide the bulk of the experimental evidence in subsequent chapters are emergent in just this sense. The high-level properties of interest, such as altruistic behavior or the investment of health by parents in their offspring, are not specifically coded for in the basic rules of the simulation. Those basic rules are generally not as low level as those of Conway's Game; out of a pragmatic concern to get things done, we usually chose middle-level behaviors, such as eating, mating and moving, as a basic vocabulary of actions for our agents. It was out of this kind of language of basic actions that more interesting patterns of behavior developed, in a way perfectly typical of much of artificial life simulation.

3.2.4 Another Definition of Simulation

Returning to the definition of simulation, we can see that one immediate benefit of Hartmann's definition is that it rules out the misleading virtual reality scenarios directly: since the human user (trainee) is a necessary ingredient, these are not computer simulations. We will nevertheless now drop the word "computer" and talk about simulation most generally, as this will help us understand the relation between the epistemology of simulation and the epistemology of experiment. Our proposed semi-formal rendition of Hartmann's mimicking account is:

Definition 3.2 *S is a simulation of P if and only if*

1. *P is a physical process (**token** or **type**)*

2. *S is a physical process (token or type)*

3. *S and P are both correctly described by a dynamical theory T containing (for S; parenthetically described for P):*

   - *an ontology of objects O_S (O_P) and types of objects $\Psi_i(x)$ ($\Phi_i(x)$)*
   - *relations between objects $\Psi_i(x_1,\ldots,x_n)$ ($\Phi_i(x_1,\ldots,x_n)$); hence, there are states of the system, s*
   - *dynamical laws of development (possibly stochastic): $f_S(s) = s'$ ($f_P(s) = s'$)*

In other words, both the simulation S and the target of the simulation P are physical processes with a common dynamical theory T. Computer simulations are then easily defined as:

Definition 3.3 *S is a computer simulation of P if and only if*

1. *it is a simulation of P*

2. *and it is a computer process or process type.*

An immediate objection to Definitions 3.2 and 3.3 might occur to you. It is symmetric: according to this, we could just as well use the sun to simulate our astrophysical programs as vice versa! This clearly won't do;

our simulations, whether computer based or not, are surely *intrinsically simpler* than their targets of study. To use a metaphor of Giere's (1999) (borrowed from Borges, 1954), if we were to construct a map of the earth on a 1:1 scale, it's true that we could more accurately measure distances using this very fine resolution map than using cruder maps, but, obviously, all the other advantages of maps would be lost. (Borges' characters start shifting domicile from reality to map!) The problems with symmetry are both practical and theoretical. Theoretically, whatever one's view may be about the nature of scientific explanation and theories, it's entirely clear that they somehow *summarize* features of the world. Computer scientists would say they *compress* information about the world. In short, they are shorter than any direct, exhaustive description of their objects.

All of this is well taken, but it doesn't follow that we need to acknowledge the point formally, within our definition. Simulations are typically constrained by both a lack of understanding of fine details of the objects of our simulations and by a lack of time to wait for the implications of fine details to filter through our simulations. Frequently, however, crude simulations are made less crude, as advances are made in both our understanding of the physical systems and in our computational capacities. If we were somehow to extend this advance in resolution power indefinitely, we might begin to approach the 1:1 scale contemplated by Giere. Admittedly, going all the way would be pointless. However, that doesn't mean that by going all the way we would no longer be dealing with a simulation; a pointless simulation remains a simulation.

3.3 Homomorphic Simulation

So, for practical and theoretical reasons, we require that our simulations *not* be as detailed as the processes we simulate. Instead we require that:[6]

Proposition 3.1 *There should exist a* **homomorphism** *h from P to S.*

Definition 3.4 *A homomorphism h from P to S is a mapping* $h : P \to S$ *such that*

1. *For every object* $x \in O_P$, $h(x) \in O_S$.

2. *For every relation* Φ, $\Phi(x_1,\ldots,x_n)$ *is true of P iff* $h(\Phi) = \Psi$ *and* $\Psi(h(x_1),\ldots,h(x_n))$ *is true of S*

---

[6]For a similar account, see Norton and Suppe (2001), although their account is somewhat cluttered with a variety of idealized, averaged and approximate models.

3. *For every state transition function f in P, $f(s) = s'$ iff $f_h(h(s)) = h(s')$ (or, for stochastic laws, the distributions over states should be identical)*

The application of homomorphisms to simulation, to be sure, should be taken with a grain of salt. That is, it is an ideal and one which we are unlikely actually to reach with non-trivial simulations. It is frequently noted in the literature that our simulations often diverge from reality in small ways and sometimes in large ways. Nearly every simulation diverges in at least this way: digital computational processes cannot exactly simulate continuous time, whereas real systems at least appear to develop in continuous time; thus, these systems support relations ("in-between times") that have no counterpart in their simulations. Nevertheless, at least for most problems, time can be discretized to a fineness where this difference does not matter. The epistemological problem is to sort out when the divergences do matter to inferences about the real systems.

Our central epistemological proposal is that simulations can be tested for adequacy by testing whether a homomorphism between the real and the virtual system holds. In the simulation literature this is called **validation**.

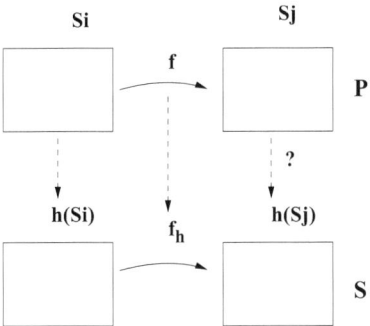

Figure 3.3: Testing for homomorphism.

3.3.1 Testing for Homomorphism (Validation)

By observing, or arranging for (by experimental intervention), the physical system P in state s_i and its subsequent transition to state s_j, we are enabled to test whether the simulation undertakes in homomorphic initial conditions the like transition (or vice versa). In Figure 3.3 this corresponds to checking that simulation S shows a transition from $h(s_i)$ to $h(s_j)$. Validation can be thought of as parallel to **confirmation**. Instead of confirming how well a theory represents reality, we are confirming how well a simulation maps reality. As such, validation comes in degrees, as there will be more or less severe tests possible for the adequacy of the mapping. Indeed, we would assert that the degrees come in the form of prior and posterior probabilities of the existence of a homomorphism, exactly as with ordinary confirmation theory, were we to allow ourselves the diversion into more traditional issues in the philosophy of scientific method.

Grimm and Railsback (2005, chap 9) present a likely account of how validation might proceed. They suggest first testing low-level submodels which describe non-emergent phenomena in the simulation and only subsequently looking at higher-level systems, including properties of the simulation that emerge from interactions between submodels. At the higher levels we are conducting simulated versions of controlled experiments (Grimm and Railsback, 2005, p. 316): "We pose alternative theories for the individual behavior as the hypotheses to be tested, implement each hypothesis in the [simulation], identify some patterns as the 'currency' [standard] for evaluating the hypotheses, and then conduct simulations that determine which hypotheses fail to reproduce the patterns." For example, the beech forest simulation was designed to reproduce both the horizontal mosaic pattern of tree stands and the vertical pattern of tree cover. But subsequently unplanned for patterns in the simulation were discovered and put to good use (Grimm and Railsback, 2005, p. 7; our emphasis):

> [The simulation] was so rich in structure and mechanism that it also produced independent predictions regarding aspects of the forest not considered at all during model development and testing. These predictions were about the age structure of the canopy, spatial aspects of this age structure, and the spatial distribution of very old and large trees. All these predictions were in good agreement with observations, considerably increasing the model's credibility. *The use of multiple patterns to design the model obviously led to a model that was structurally realistic.*

Other than the fact that this procedure is dealing with a computer simulation rather than directly with an ecological theory, there is no interesting methodological difference between this and standard theory testing. A rich, multi-patterned simulation offers a variety of opportunities for testing its conformity to the target process. And just as in standard confirmation

CHAPTER 3. SIMULATION AS EXPERIMENTATION 67

theory (Franklin, 1986, pp. 123-129), the more varied the predictions of a simulation, or the submodels used to make them, that are tested and confirmed against reality, the greater our confidence that the simulation indeed maps that reality. With a sufficient variety of such tests, testing diverse transitions under diverse conditions, we may well be able to conclude that the simulation is, or is not, homomorphic, either approximately or exactly.

The existence of an approximate homomorphism is *crucial*: it underwrites the relevance of the simulation for the system being simulated and, in particular, its use both for explaining events in the real world and in predicting them.

As noted, homomorphisms may exist at a variety of levels of resolution. The level of resolution of the homomorphic simulation depends upon two major points:

1. How well do we (think we) understand *P*? How detailed a theory do we have to test?

2. Pragmatic constraints upon our simulation (e.g., how much time can we spend waiting).

The levels of potential simulation may lie upon one-another Shrek-like, as in an onion (see Figure 3.4):

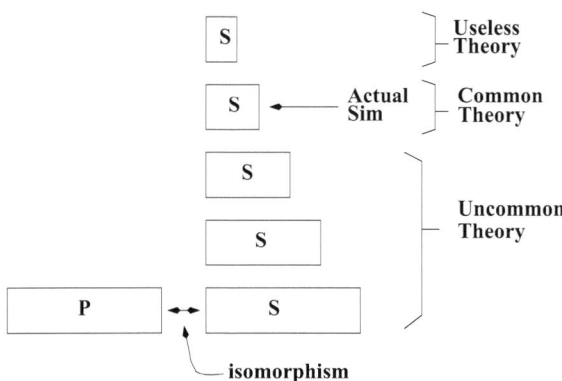

Figure 3.4: The simulation onion. (Smaller layers indicate fewer details, but greater generality.)

At the top-level is a simulation with such a small ontology that nothing useful can be simulated. Below the actual simulation are more detailed potential simulations which are unused for such reasons as: the theory describing such detail has not been invented; the simulations at that level of detail are impractical; the level of detail describes events of no interest to us.[7]

3.4 Simulations as Experiments

Many people have been attracted to the idea that simulations have no empirical side to them and, in particular, that they are basically revved-up thought experiments. Oreskes and others claim that simulations cannot be used to acquire *any* empirical knowledge about the world, directly or indirectly (e.g., Oreskes et al., 1994, Axelrod, 1997, Di Paolo et al., 2000). Rather, simulations are limited to extending our understanding of the theories being simulated, by exploring their deductive consequences. Di Paolo, Noble and Bullock (2000), for example, examine the Hinton and Nowlan (1987) simulation study of the Baldwin effect — the acceleration of genetic evolution via learning by individuals in a population (Baldwin, 1896). Prior to that study, the Baldwin effect had been given little attention; it sounded too much like a Lamarckian process and the mechanism for fixing learned behavior genetically was not understood. The simulation of Hinton and Nowlan (1987) changed that by producing a plain, easily inspected mechanism, which demonstrably exhibited the Baldwin effect. Di Paolo et al. (2000) argue that this is essentially a theoretical, deductive use of simulation, making plain what was implicit in the theory. While it is clear that simulations can be used to explore the deductive consequences of theories, it is not clear that that is the only role they may have in empirical science. Nor is it clear that Hinton and Nowlan's *mechanism* was in any sense implicit in Baldwin's theory, as it would have to be were it but a deductive consequence of that theory.

We now proceed to argue that simulations have potentially every role that experiments may have in empirical science.

3.4.1 A Comparison with Real Experiments

To further our claim that simulation studies share epistemology with traditional scientific experiments, we can consider Allan Franklin's experimental strategies (Franklin, 1990). Franklin emphasizes that his strategies

---

[7]In other words, this kind of account is very far from requiring the "perfect mimesis" of **isomorphism** that Winsberg (2003, p. 116) claims is implied.

CHAPTER 3. SIMULATION AS EXPERIMENTATION 69

are neither exclusive of other strategies nor exhaustive. Nevertheless, they provide a good indication of what happens in physical experiments; we annotate the list with reference to simulation studies (Franklin, 1990, p. 104):[8]

1. Experimental checks and calibration, in which apparatus reproduces the known phenomena

2. Reproducing artifacts that are known in advance to be present
   Regarding 1 and 2, reproducing known phenomena is a standard check of adequacy in simulation studies.

3. Intervention, in which the experimenter manipulates the object under observation
   The relative ease of manipulating simulations is one of their key advantages in experimental studies.

4. Independent confirmation using different experiments
   In simulation research there is always an opportunity to test very different kinds of initial conditions, and sometimes an opportunity to test the operation of distinct subprocesses (Grimm and Railsback, 2005, Grimm et al., 2005). Replication of simulation results using distinct simulations is also a possibility (e.g., Axtell et al., 1996, Edmonds and Hales, 2005).

5. Elimination of plausible sources of error and alternative explanations of the result
   These are activities integral to both **verifying** *and validating simulations.*

6. Using the results themselves to argue for their validity
   By this Franklin meant that an experiment may create results which are highly unlikely to be artifacts of the measurement process or experimental procedure and so by themselves support the claim that they reflect an external reality. Similarly, simulation results may likewise be determined to be highly unlikely to be due to bugs, not just because of steps in the verification process, but also because of the results themselves.

7. Using an independently well confirmed theory of the phenomena to explain the results
   This is one leg of our triangle of Figure 3.5 below.

---

[8]We have corrected Franklin's "corroboration" with "confirmation".

8. Using an apparatus based on well confirmed theory
   The apparatus here is the simulation and associated software; verification is part of the process of justifying the claim that it is based on well confirmed theory.

9. Using statistical arguments
   It has been frequently remarked that the use of, or rather the need for, automated data analysis and data visualization techniques is a striking feature of simulation research. Epstein and Axtell (1996), for example, employ a variety of graphics to good effect.

Clearly, at a phenomenological level, simulation research is very akin to traditional experimental research. But this does not demonstrate that at an "epistemological level" they are again alike.

3.4.2 The Epistemology of Simulation

There are two acknowledged steps to justifying claims that a simulation is informative of the real world:

Verification: Determine whether the simulation correctly implements the theory being investigated,

- e.g., by performing design verification, debugging the simulation, and consistency checks.

Validation: Determine whether the simulation as implemented conforms to the physical target process.

- This is testing for the existence of a homomorphism, by comparing simulation results with the target process and vice versa.

These steps are portrayed graphically in Figure 3.5. A theory T has been developed for the type of physical process P.[9] A range of **token** processes, $P_1, \ldots, P_n$ instantiate that **type**. And a simulation has been developed for those processes and/or the process type, leading to simulation processes $S_1, \ldots, S_m$. We can think of the computer program used to launch the simulation processes as a process type S, not depicted in the figure. This situation presents us with a triangle with three legs of possible justificatory test for the relevance of simulation S to its target P: whether the theory T represents

---

[9]Incidentally, we are favorably inclined towards the semantic interpretation of scientific theories (Suppe, 1977), but nothing in our account hangs directly upon that.

CHAPTER 3. SIMULATION AS EXPERIMENTATION

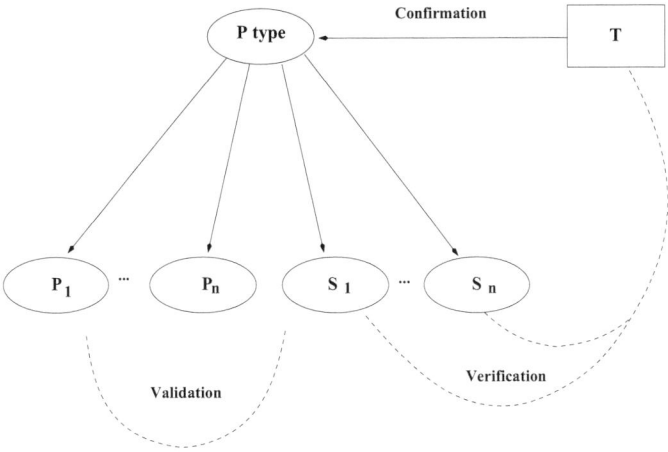

Figure 3.5: The verification, validation and confirmation triangle.

the reality P (confirmation); whether S properly implements T (verification); whether S corresponds to P (validation). Each test can be carried out independently of the others. Once any two tests have been conducted, and assuming their outcomes are not in dispute, then the third test becomes irrelevant, since we then know everything there is to know about the relations between T, P and S. This explains a range of observations previously made about simulation. Given verification, S can be manipulated to investigate the implications of T (this is the "S as thought experiment" above). Again, given verification, determining that there is a homomorphism between P and S is tantamount to confirmation; given validation, S can be used in exploratory theorizing as well as predicting the consequences of intervening in P systems; given confirmation, failures of correspondence between P and S indicate verification failures. When any two justificatory steps have been successful, the S_k are just as much instantiations of P as are the P_i: they all provide supervenience bases for that type of physical process.

3.4.3 Experiments as Simulations

An obvious rebuttal to the claim of epistemological sameness between experiments and simulations is: "Unlike simulations, when you're testing the real-world at least you know what you're testing is *real*! You can't be testing the wrong thing!" Morgan (2002), for example, claims there is an inferential gap between simulation and reality that doesn't exist between real-world experiments and real-world target systems. This is a seductive thought, but it is wrong. The inferential gap is always there in any scientific study. It is nearly an everyday occurrence to hear about some medical study in which the experimental groups turn out to be unrepresentative of some target population. And it is an old joke that experimental psychology has accumulated a large body of evidence about the psychology of college students.

We previously observed that, following our Definition 3.2, the targets of our computer simulations — the physical processes in the world — might be construed as simulations (but not *computer* simulations) of our computer processes, even if that is not pragmatic. The general point is that some (non-computer) physical processes may be used to simulate others. For example, scale models built with clay and water are used to assess water flow and tidal action; or, wind tunnels are used to assess aerodynamic flows. Again, of course, the simpler and smaller processes are generally said to be simulations of the more complex and larger processes which are the targets of investigation. Which is the simulation is, in the first instance, driven by which process is wanting to be understood. But other factors play a role, including accessibility, ease of intervening in the process, and the ethics of intervening in the process. If there is a single process, and not a process *type*, of interest, and if it is accessible, etc., then there may be no recourse to a simulation: a simple intervention in the target process may be attempted. There is then no inferential gap, because the studied system is also the target system. Engineering applications and most treatments by medical doctors are of this type. If our interest spans an entire type of process, and if we are not looking for a specific outcome other than learning about that type, then a direct examination of all instances of the type is unlikely to be possible. In that case, we shall have to simulate the process type of interest with another that is more accessible. For example, we might use a sample of adult humans in Australia today to simulate the category of all adult humans across all of time and space. This is the common practice in medical research. Similarly, we might be interested in physical conditions immediately after the Big Bang, but have little prospect of directly measuring them; instead, we might simulate such conditions using high-intensity collisions of subatomic particles. Again, Galileo famously tested his telescopes on terrestrial objects to gain support for his inferences concerning celestial objects (Chalmers, 1982, p. 72). While there are many complications and nuances to each of these stories, in all these cases and all such experiments we are using one physical process, or type of process, to simulate another. And, in all such experiments, there is the same potential for things to go wrong, for the experiment to be uninformative, because the experimental subject fails to simulate the target subject, because the homomorphism between reality and experimental process fails.

In short, "studying proximate systems as stand-ins for target systems of interest... pervades all science" (Frigg and Reiss, 2009). The epistemology of computer simulation is the same as the epistemology of experimentation for the simple reason that all experiments are simulations and computer simulation experiments are just a special type.

3.4.4 Special Epistemology

Regardless of our arguments and proposed interpretation of the epistemology of simulation, various philosophers of science have claimed that computer simulation is a new methodology demanding a new epistemology (e.g., Winsberg, 1999, 2003, Humphreys, 2004, p. 54). Computer simulation is certainly a new methodology, involving new tools and techniques. Expertise in experimental physics is hardly interchangeable with expertise in physics simulation. But what additional reasons have been advanced for demanding a new epistemology? Some of the features of simulation said to require new epistemological thinking are:

- *Visualization.* Coupling computer simulations with visualization, including animations, is very common; indeed, computer simulation is being used specifically to create artistic animations (e.g., McCormack, 2005). Humphreys (2004, pp. 111-114) seems to think that the use of visualization is a *defining* characteristic of simulation, which is clearly going too far. Many simulations have been done without any graphical displays being generated. Regardless, Frigg and Reiss (2009) are surely right that the importance of visualization in coping with massive amounts of data is a property that simulation *shares* with experimentation.

- *Approximation.* All computer simulations, short of simulations actually isomorphic to the target process, are, of course, approximate. But this again is no unique property of computer simulation. There are many examples of the use of physical experiments which are known to distort the properties of target systems. Wind tunnels are used to investigate aerodynamics, however their walls introduce "unnatural" turbulence which can affect the process under study (Norton and Suppe, 2001). Some such distortion is true of any scale model and, more generally, of any physical system not strictly identical, or isomorphic, to the target system.

- *Discretization.* Something which may well be distinctive about computer simulations is that they are housed within discrete von Neumann machines. We've already mentioned that time must be represented as a sequence of time steps, whereas real processes do not step through time. More generally, any state variable must be represented with some finite degree of precision, implying misrepresentation of real numbers. The difficulties associated with discrete approximations representing continuous systems are often exaggerated. Chaotic systems, to take a common example, are famously characterized by a sensitivity to arbitrarily small variations in their initial conditions, so even within an arbitrarily small degree of precision adopted by a discrete representation a chaotic system's behavior may display sensitivity. These are sensitivities to which that discrete representation must remain blind. But that doesn't stop those experimentally investigating chaotic systems from the near exclusive use of digital computers to run their chaotic simulations! The chaotic behaviors of interest to them must nevertheless be exhibited at the discrete levels of resolution which they adopt with their simulations. Typically, the variations beneath that level of resolution are simply variations on a theme already reflected in their discrete simulations. In any case, the attempt to avoid such sensitive response to initial conditions by using analog computers would be pointless: there is always a limit to the degree of precision available to us, whether we use digital or analog equipment.

  This kind of potential epistemological concern generalizes to any experimental apparatus: although in a real-world experiment the experimental and target processes may well both be continuous processes, the experimenter will have no way of taking advantage of that (in either manipulation or observation) beyond some finite degree of error. In particular, noise in experimental measurements will always distort the experimental results. Such distortions may indeed undermine the relevance of the experiments for the desired confirmatory conclusions. This means the epistemological difficulties for discrete approximations apply equally to classical scientific experiments and to computer simulation experiments.

- *Calibration.* Typically simulations have various parameters that need

CHAPTER 3. SIMULATION AS EXPERIMENTATION

to be calibrated so as to reproduce known phenomena of the real system. For example, in a simulation of the evolution of group selection the strength of altruistic behavior needed to be adjusted to produce a stable population of altruists (Appalanaidu, 2007). This might suggest that you can get whatever result you want by recalibrating your simulation. It's implausible that science constructs reality according to its wishes, despite the more extreme views of social constructivists, but it's far more plausible that computer scientists can construct virtual reality according to their wishes. We accept that there is some danger here; the flexibility of universal computation can cover many faults of theory, if allowed to do so. However, we again assert the parallelism between simulation study and experimental study: given the validation of structural properties of a simulation, the calibration of parameters of the simulation can only push the results so far and not infinitely far. Such calibration serves the identical purpose with calibration in physical experiment, that of finding the settings which support previously observed measurements of a target system under given initial conditions, and so supporting the claim that measurements under new conditions will be informative.

3.5 Conclusion

There remains a residual reflexive skepticism about the epistemological worth of computer simulation experiments, which is more often expressed about contentious science, such as the science of anthropogenic climate change or artificial life. But the difficulties which those studies encounter, while real enough, are no different in principle from those encountered in non-controversial simulation science, such as in astrophysics or meteorology. And none of these are different in principle from the difficulties encountered in "ordinary" experimental studies of medicine, psychology, chemistry and physics, which use real-world experimental models to simulate real-world target systems.

So, the experimentation with computer simulations that has become a prominent feature across the sciences is more than experimentation in name only. It is full-blooded experimentation. It carries problems, techniques and methods which are clearly new, such as software debugging methods. It carries with it problems, techniques and methods which are old as well, such as figuring out which statistics to capture to obtain an informative view of what is happening. None of the issues raised to this point actually identify any telling limitation on computer simulation or the need for any new epistemology. The difficulties with sorting out the epistemology of experimental science are not yet adequately resolved; but there is no reason to believe that that epistemology won't have rich enough resources to accommodate what scientists are today doing with their computers.

The limits of computer simulation are, thus far, the limits of Turing computation. We know some of what lies beyond those limits, what has been called hypercomputations (Copeland, 2002), computations which, for example, infinitary machines can deal with. Despite the pessimism of Dreyfus, Penrose and Humphreys (Dreyfus, 1992, Penrose, 1999, Freedman and Humphreys, 1999), and many others, however, we have been given no reason to believe that human mental capacities are beyond the capabilities of Turing computability. In consequence, the prospects for the "ultimate" simulation, that of the human brain — completing the practical goal of producing a genuine artificial intelligence, are very real, if also rather distant.

Chapter 4

Evolutionary Artificial Life

Artificial life replicates living processes in artificial media, most notably perhaps in chemistry, with Craig Venter having announced the first artificially constructed, living cell in May 2010 (Sample, 2010). We are here, however, focused on computer media: artificial life within virtual worlds. Virtual life provides the maximal flexibility within which to investigate the processes of life, and, as the namer of the field famously emphasized, it allows us to investigate life both as it is and as it *could be* (Langton, 1989). Despite the latter possibility — the speculative investigation of counterfactual worlds and counterfactual life forms — it is ALife's potential for investigating the actual world and its actual denizens counterfactually that animates us here.

While the term 'Artificial Life' is just two decades old, ALife research began earlier. As early as 1911, Leduc experimented with colloidal solutions that could emulate metabolic functions, mitosis, and other activities associated with life in a limited way (Keller, 2003). Later, John von Neumann (1951), inspired by Stanislaw Ulam, developed the idea of **cellular automata** (CAs). CAs are regular lattices of cells each of which operate the same finite state machine, taking as input the states of its neighboring cells and performing some transition to a new state. The best known example of a CA is Conway's Game of Life. That, as we saw in Chapter 2, consists of a few simple state transition rules replicated throughout a two-dimensional array. The Game of Life produces self-replicating patterns (gliders) as we saw. Von Neumann was the first to produce self-replicating structures in CAs, although they were rather more complicated than Conway's.

All of the simulations we describe here can be viewed as CAs, even though they are described in terms of agents, their environments and their behavior, which includes moving from cell to cell. The CA view simply redescribes, for example, an agent moving from a cell to a neighboring cell as two adjacent cells changing state in concert with each other. The CA view of simulation can be generalized in various ways, for example from two dimensions to three, or even from computer simulation to physical reality, with Wolfram (2002) suggesting that the entire universe is nothing but a gigantic cellular automaton. However, we will not be pursuing the CA view of our simulations because it is both less natural and far more complicated.

As we are studying life as it *is*, this requires us to look at life as it *evolves*. The idea that evolution could contribute to machine learning occurred to people early on. Oliver Selfridge included genetic operators and selection in his program *Pandemonium*, for example (Selfridge, 1959). In the 1960s others began investigating genetic algorithms (Holland, 1975), the evolution of finite state machines (Fogel et al., 1966) and the evolutionary optimization of real-valued parameters (Rechenberg, 1971), with these efforts proceeding independently. As Selfridge had already discovered, however, evolutionary processes are computation-intensive, and none of these approaches caught on before the PC revolution of the 1980s, which also reignited interest in other computation-intensive techniques, including **artificial neural networks (ANNs)** and Bayesian modeling.

Around the same time that artificial evolution was re-emerging, and its different strands discovering one another, ALife was emerging from an obscure investigation of cellular automata into a self-aware program of investigating life. Chris Langton brought this kind of work together under the banner of 'Artificial Life' at the *International Conference on the Synthesis and Simulation of Living Systems*, also known as Artificial Life I (Langton, 1989). Since then, ALife has grown into an interdisciplinary network of ideas from physics, philosophy, social science, economics and more (Adami, 1998). Importantly, the value of **evolutionary algorithms** for computer optimization and machine learning has lead to ever increasing interest within ALife in simulating evolutionary processes.

In the rest of this chapter we outline the main design elements that figure in our simulation and their intellectual history, including the confluence of evolutionary algorithms and artificial life techniques. We will also contrast our simulation methods with related, but distinct, kinds of artificial life methods, such as the social simulation of Gaylord and D'Andria (1998) and individual-based modeling in ecology and epidemiology.

4.1 Simulated Evolution

4.1.1 Genetic Algorithms

The genetic algorithms (GAs) developed by John Holland in the 1960s were among the first, and remain among the best known, evolutionary algorithms. GAs are detailed in Holland's monograph *Adaptation in Natural and Artificial Systems* (Holland, 1975). As we mentioned above, techniques that are similar in spirit, likewise incorporating the basic elements of evolutionary processes, include evolutionary strategies (Rechenberg, 1971) and evolutionary programming (Fogel et al., 1966); towards the end of the 1980s these became collectively known as evolutionary algorithms. GAs became perhaps the most influential of these approaches through their widespread adoption as a general-purpose optimization technique, which in turn led to growing interest and use by ALife researchers.

GAs are algorithms used to search for "good enough" solutions to well-defined problems and, in particular, problems whose candidate solutions can be readily measured for distance from an acceptable solution using a "fitness function" (Goldberg, 1989). A GA operates over a population of candidates, which originally (in Holland's account) had to be bit strings, but which in current usage can be any kind of digital representation.[1]

Like biological evolution, genetic algorithms have the three necessary ingredients for evolution (see § 1.2.1): heritability, or reproduction with inheritance; variation, through reproductive crossover and mutation; and selection. GA mutation is a simplification of mutation in natural evolution, acting as an artificial analogue to natural copying errors, usually achieved by flipping bits randomly (with a low probability) at reproduction. However, reproduction and selection have been altered in order to produce a better search algorithm. In particular, these operators have been modified to operate in accord with a "fitness" function (or, objective function), where fitness is inversely proportional to the measured distance to a satisfactory solution to a given problem. The better a candidate solution is according to this fitness function (the closer it is to a solution), the more likely it is to reproduce and so the more probable components of its genome will be represented in future generations.

While GAs are effective and simple tools for computerized optimiza-

---

[1] Although any representation within a standard digital computer can be thought of as a bit string, it does not follow that any such representation should be treated as a generic, undifferentiated bit string. For example, in a GA bit string representation of real numbers a bit mutation of the exponent will have far greater impact than a bit mutation of the mantissa.

tion, they are not intended to be close simulations of any biological evolutionary process. Aside from the considerable simplifications in genetic operators such as mutation, GAs commonly represent individuals simply as chromosomes without any bodies! There is little or no phenotype or behavior. For example, reproduction is typically driven entirely by the demands of a fitness function applied to the chromosome, rather than adaptedness to an environment; there is no environment (Mitchell and Forrest, 1994). Naturally, GAs are not used to study biological evolution (Wilson, 1989). Their contribution has been instead to inspire other, more realistic, simulations which can tell us something about biological evolution.

4.1.2 Simulating Evolution

An early and famous case of simulated natural evolution is that of *Tierra*, developed by Tom Ray (1991). *Tierra*, in contrast with GAs, has individuals with bodies that behave in an environment. However, Ray took seriously Langton's suggestion that ALife might investigate life as it *could be:* the bodies are computer instructions, their environment is computer memory, and the energy required for "metabolism" is CPU time. Once a computer organism is born, it behaves by copying words of computer memory to new locations, and perhaps thereby reproducing itself; it also ages and dies (i.e., its memory cells are reclaimed by the system after awhile). Organisms in *Tierra* are run for a fixed number of CPU cycles; those that require fewer machine instructions to reproduce themselves spread copies through the computer memory, outcompeting and displacing organisms that need more instructions to make copies.

As a biologist, Ray's intention was to mimic evolution in nature, without simulating anything like an actual evolutionary history. His most noted result was that, beginning with simple reproducers, *Tierra* evolved even simpler reproducers, something like viruses that infected larger reproducing chunks of code that would copy both themselves and their parasitic invaders. Indeed, Ray's simulation was able to produce "parasites, immune hosts, immunity eschewing parasites, symbionts, cheaters, super-parasites, etc." without being seeded with any of these (Adami, 1998). His efforts inspired the creation of many further simulations exploring evolutionary processes and the origins of evolution through similar means, such as *Avida* (Ofria and Wilke, 2004) and *Amoeba* (Pargellis, 1996). It is a remoter inspiration also of attempts at more biologically oriented evolutionary simulations in ALife.

As we mentioned in Chapter 3, a particularly influential ALife simulation of evolutionary processes was Hinton and Nowlan's (1987) simulation establishing the possibility of the Baldwin effect — that is, the possibility that learned traits can become fixed through evolution. It was an early demonstration of how computer simulation could directly support a biological theory.

4.1.3 Evolving Psychology

Evolutionary psychology aims to understand the mind and the universals of human behavior by reference to evolutionary mechanisms. Many of the theses of evolutionary psychology generalize beyond humans to other species. It is this generality that allows evolutionary psychology and current ALife to intersect. Within evolutionary psychology, this intersection covers the general models of behavior; within ALife, it covers evolutionary agent-based simulations, including simulations such as Hinton and Nowlan's (1987) and those we present here.

One might suppose that evolutionary psychology has more in common with artificial intelligence than ALife. However, AI is not directly concerned with understanding the evolution of cognition, while ALife is. AI's interest in evolution, aside from its potential for providing machine learning within individuals (which does not correspond to natural evolution), is in finding inspiration or means for engineering intelligences that have not evolved. Evolutionary ALife aims primarily at simulating evolving populations in order to solve practical problems, to understand evolved populations and ecosystems, or to understand the evolutionary processes which brought them about. Insofar as the problems or their populations engage in complex social dynamics, evolutionary ALife must engage in the evolution of social behavior and, with it, the evolution of cognition.

The central ideas of evolutionary psychology have found their way into ALife — ideas such as kin selection, biological altruism and parental investment. For their part, many ALife simulations have the potential to contribute to evolutionary psychology. In addition to Hinton and Nowlan's (1987) simulation of the Baldwin effect, there are those of Gintis (2000), Jaffe (2002), Mitteldorf (2002) and our own simulations. So far, there is a good deal more opportunity for collaboration between ALife and evolutionary psychology than practice, but we anticipate rapid change in their relationship.

4.2 Individuals, Agents and their Societies

ALife simulations often have large populations of individuals that observe their environment, act under their own power, and interact with each other and their environment. Such individuals are called *agents* and the simulations are called agent-based models (ABMs). Agents are unlike the bit strings of a GA, or the program "organisms" of Tierra, since they are not only capable of surviving and coping with some environment, but they often also make decisions and change their behavior throughout their lives.

Effectively identical simulation methods have been adopted widely in the last two decades in the social sciences, epidemiology and in ecology. Ecologists call their artificial life simulations individual-based modeling (IBM) (Grimm, 1999), in order to contrast them with the more traditional simulation technique of population-based modeling. Population-level models apply sets of equations to simulate dynamic systems. The equations describe the state changes of the system over time, as a single entity. For example, the Lotka-Volterra predator-prey equation might be used to simulate the relationship between populations of foxes and rabbits (Volterra, 1931). The individual-based modeling alternative is to design representations of foxes and rabbits which include attack, defense and reproductive mechanisms, populate a common environment with an initial population of both kinds of individual and let them go. In such a case, if the simulation is done correctly, the Lotka-Volterra equation will *emerge* as the correct description of the IBM's dynamics. The obvious cost of the IBM approach is that the simulation is considerably more complex. If all you actually want is a Lotka-Volterra model, then the population-level simulation will be far easier to set up. The benefits, however, can far outweigh the cost. Instead of treating every individual in the population as identical, as the population-level model must, IBMs allow the introduction of varying abilities to attack, defend and reproduce through the two populations. Real populations, of course, exhibit considerable variability in these traits, and that variability impacts not just on evolutionary processes (which demand them!) but also on the population dynamics themselves, especially when the environment being simulated is itself heterogeneous. In short, the more realistically we wish to model populations, the more pressure there is to move from the population level to the individual level. This (together with continuing improvements in computer technology) explains the ever growing popularity of ALife simulation, even if largely in the guise of individual- and agent-based modeling, or again as social simulation in the social sciences.

We will soon describe the software architecture we use to introduce, maintain and evolve populations of individually diverse agents. But first we will set it in context by reviewing some of the more prominent simulations in the literature.

4.3 Some Ancestral Simulations

The Iterated Prisoner's Dilemma (IPD) (see §1.4.1) has acquired a kind of iconic status for those interested in the evolution of behavior. It was particularly the emergence of Anatol Rapoport's strategy tit-for-tat as the winner of Axelrod and Hamilton's 1981 IPD tournament that excited the imagination. Axelrod (1984) went on to show that tit-for-tat would evolve given a population of random initial strategies playing IPD, and this has repeatedly been trumpeted as a model for the evolution of cooperation and altruism.

Axelrod's computer simulation was a genetic algorithm with 64-bit chromosomes capable of selecting an action conditioned upon the prior three rounds of the IPD (i.e., three actions each for the player and its opponent). Chromosomes (strategies) which accumulate larger rewards in IPD were then probabilistically preferred for reproduction, and so over-represented in the next generation. Axelrod's GA also incorporated the usual sources of variation, in chromosomal crossover and occasional mutation.

It's easy to see how tit-for-tat may outcompete other strategies. When playing against always-defect, tit-for-tat mimics always-defect after the first turn. However, when tit-for-tat plays against tit-for-tat, it acts like always-cooperate. So, in a population composed of some proportion of each, it is easy for tit-for-tat to outcompete always-defect, even in cases when always-defect is the most common strategy. tit-for-tat not only evolves, but can also be an evolutionarily stable strategy, resisting invasion by defectors. Cooperators will do as well in a pure tit-for-tat environment, of course, but if both cooperators and defectors arise occasionally by mutation, then they will not, as they will be bled by the defectors while tit-for-taters will not.

The evolution of tit-for-tat and its stability suggest that tit-for-tat may have been an important behavioral strategy during evolution, under the interpretation of the IPD payoffs as payoffs to fitness. The "niceness" of tit-for-tat (its initial cooperativeness), combined with its readiness to retaliate against defections, may well contribute to evolutionary environments that are more prone to support altruistic behavior, for example by helping to generate the diversity of groups required by group selection (see §5.1), even though tit-for-tat itself easily evolves within homogeneous populations.

Subsequently, Nowak and Sigmund (1993) discovered another robust strategy, which they called Pavlov, in which both players cooperate if and only if they chose the same action (whether cooperation or defection) in the last round. Pavlov was so called because of its reflex-like reaction to payoffs: it changes behavior only if it receives a low payoff, i.e., whenever the other player defects (hence another name for this strategy is win-stay, lose-shift). Nowak and Sigmund found that Pavlov evolves as the dominant strategy under more realistic conditions than Axelrod used, namely where players can make occasional mistakes and where **mixed strategies** are allowed, even though Pavlov itself is a deterministic strategy. While Pavlov is unable to invade populations of defectors, it can invade tit-for-tat populations, which in turn can invade defector populations.

The simulations of Axelrod and Nowak and Sigmund demonstrated the great potential for simulation to contribute to the evolutionary understanding of social behavior and have become some of the most widely cited work in recent social science and evolutionary psychology. Nevertheless, as we mentioned in Chapter 1, there is some reason to be skeptical of the enthusiasm with which IPD simulations have been received. There is, after all, only the appearance of cooperation in tit-for-tat without the reality of mutual goal-seeking. Nor is there anything like altruism there, since there is no cost to the individual's self-interest: on the contrary, in the circumstances in which it evolves, tit-for-tat is maximizing self-interest. A far more appropriate model for the evolution of altruism is the evolution of *suicide*. If the act of suicide can be arranged to benefit others, then it is obviously altruistic in almost all circumstances. The only question is whether it can be made to evolve, since on the face of it it should only evolve *away*.

Epstein and Axtell's *Sugarscape* followed ten years after Axelrod's simulation, bringing the ideas of artificial life to bear on the concerns of social science in the "social simulation" of artificial societies (Epstein and Axtell, 1996). The design of *Sugarscape* is very similar to the design of the simulations in this book, but the authors were motivated by very different questions. *Sugarscape* is a two dimensional board, a lattice of 50 by 50 cells, in which agents and various other resources live. The agents are able to move and see in the four directions of the compass for varying distances. The board is sprinkled with sugar, which replenishes the energy consumed by motion and metabolism. If the agent's energy level drops to 0, the agent dies.

CHAPTER 4. EVOLUTIONARY ARTIFICIAL LIFE

With this basic design Epstein and Axtell were able to investigate several different phenomena. We will briefly describe two: the first, a simple simulation that exhibits emergence; the second, an investigation of disease transmission that shares traits with our own simulations of **aging** and group selection.

In the first of these simulations, the north-east and south-west corners of the board are home to alternating seasons so that when one is rich in food the other is in drought. In this environment, the agents move together in periodic and diagonal wave-fronts, migrating seasonally between these two corners of their square world. They move as if collectively seeking the corner currently in bloom, excepting some agents with low vision who never migrate and instead wait for the next spring, whom Epstein and Axtell call 'bear-like hibernators'. This wave of migration can not be explained from the rules governing individual agents alone. Agents are not programmed to seek corners in springtime; they are not even capable of diagonal vision or motion. Hence, their diagonal migrations are not directly explained by the rules governing individual agents, but are emergent from them in the same way that gliders and glider guns emerge from Conway's Game of Life.

The second simulation is of disease transmission. Epstein and Axtell represent diseases as bit strings of a certain length. Each agent has an immune system, which is another bit string, several times longer than the longest disease. Agents will be immune to a disease if the disease bit string is a substring of the immune system, and otherwise uninfected agents that are near infected agents will catch the disease. After catching a disease, an agent's immune system will try to immunize the agent against it by finding a substring that most closely matches the disease (based on the **Hamming distance** between them, i.e., the number of bit flips required to turn one into the other). The agent builds immunity by modifying this substring one bit at a time (one bit per cycle) until it matches the disease's bit string. While the agent is building its immunity to a disease, it remains infected: in the simulation, this translates to a higher metabolic rate and the potential to spread the disease to other agents.

Epstein and Axtell found the entire agent population would become disease-free when the simulation contained only a small number of diseases. In contrast, when the number of diseases increased — in particular, when the sum of the bit string lengths of all diseases exceeded the size of the immune system bit string of a single agent — diseases persisted indefinitely.

Sugarscape served the purpose of advertizing the potential of ABMs for examining economics, social behavior, cultural change and epidemiology. Its extreme simplicity was both virtue and vice, virtuously demonstrating the flexibility and power of the simulation methods, while viciously limiting its ability to realistically portray or predict any actual social system or disease. The inspiration it provided, however, has served its purpose, and ABMs now include some of the most useful predictive models for disease transmission (e.g., Longini et al., 2004, Ferguson et al., 2005).

	IPD Simulations	Sugarscape
Analysis of social behavior	✔	✔
Ethical implications	✔	✘
Uses evolution	✔	Sometimes
Evolutionary psychological concepts	✔	✘
Implicit fitness function	✘	✔
Agents and food	✘	✔
Environmental interaction	✘	✔

Table 4.1: Properties of our own simulation shared by the IPD and *Sugarscape* simulations (4) and those that are not (5).

We employ a similar disease transmission model in some of the simulations in Chapter 5, particularly as a means of examining the effects of co-evolution, where disease and host respond to each other's presence genetically. In Table 4.1, we summarize the similarities between the IPD and *Sugarscape* simulations and our own.

4.4 A Simulation Environment

We now present the main features of our own simulations. The design elements and architecture we describe here are not used in all of our simulations discussed in the next two chapters, but do provide a common basis for most of them. Any variations from the details we give here are, of course, indicated during the subsequent discussions of those experiments.

The environment within which the artificial agents behave and reproduce is a critical feature of any simulation. As with the vast majority of ABMs and IBMs our simulations here maintain the simplifying fiction that the agents and their environmental "furniture" are entirely distinct and, in particular, do not exchange matter. There is, however, an exchange of energy, with agents consuming food and, usually, returning their energy to the

CHAPTER 4. EVOLUTIONARY ARTIFICIAL LIFE

environment upon death. More realistic relations between agents and their environments are considered in the simulations of Dorin and Korb (2007), Dorin et al. (2008).

4.4.1 The Simulation World

The activity of the simulation takes place on a board — a grid of $n \times n$ cells, where n varies across experiments. In some experiments each cell can take an unlimited number of entities, in others cells are limited to a single agent. The world is sometimes bounded at the edges, so that agents cannot move beyond; in others the opposing edges connect, creating toroidal topologies (donuts).

4.4.2 Time

There are two units of time in our simulations: cycles and epochs. A cycle is a pass through all the agents in the simulation, in which each gets the chance to act. A random order is used each cycle to avoid ordering effects. An epoch is a period of cycles during which the system collects statistics.

Cycles are often implemented in two passes. The first pass lets agents observe a common state of world when choosing its action, while the second allows them to act. Using two passes is of particular value in simulations that have seasonal droughts: if agents consume all the food available during spring, then agents towards the tail of a single-pass cycle, faced with no food, would confuse this short gap in food supply with a longer-term drought. A much more complicated alternative is to implement multi-threaded (simultaneously active) agent simulations.

4.4.3 Food

Food is used to fuel the agents and has the following properties: it can occupy a cell; it can be consumed, which removes the food from the cell; and, in some simulations, it has a finite lifetime even if not eaten. Each piece of food also has an associated health, or energy,[2] that is used — and only used — to boost an agent's health when the agent eats it.

Food is generated by the system using the food distribution function (*fdf*) mapping the current time to an amount of food. In some cases, the function is sinusoidal (i.e., seasonal) and may have a period and amplitude implying droughts, while in others it is constant.

---
[2] We make no distinction between health and energy in our simulations here.

4.5 Agents

The agents can have quite divergent properties and behaviors, depending on the experiment. Here we describe some of the more common properties.

4.5.1 Birth, Age and Death

Agents are predominately created by sexual or asexual reproduction, although at the beginning of a simulation there is normally a randomly generated population, dominated by unfit types. In the **co-evolution** simulations, a species may be generated spontaneously, ensuring parasitic species do not die out.

An agent's age is measured in cycles — specifically, the number of cycles that have passed since the agent's birth. Any gestation period does not count towards an agent's age. There are a number of ways an agent may die. Each agent is given a maximum age at birth, which the system generates according to Gaussian simulation parameters. An agent may well die prior to its maximum age, which usually occurs when the agent's health is exhausted. In some simulations an agent may also die accidentally or by committing suicide. When an agent dies, the system removes it from the board.

4.5.2 Health and Utility

A simple numerical variable tracks an agent's health. An agent will begins life with some initial health value which will then change in response to actions it or other agents perform. If an agent's health falls to zero, it dies.

Initial Health. Initial health is the sum of initial investments of health by each parent. In some experiments the parental investments are of a fixed size, while in others they evolve. For agents at the start of a simulation, the system assigns an initial health that is normally several times higher than the average parental investment. This ensures that some of the population can survive the early stages of the simulations, which, given the randomly generated and highly diverse gene pool, display stronger selection pressure than at any other time.

Health Effects. As the agent performs actions, its health may fall from health costs associated with the actions. Eating transfers the food's health store to the agent, minus the cost of eating itself. In some experiments, agents derive less health from food than the food makes available — i.e., the food is poisonous. An agent may also boost its health by resting, when the action is available — though this health boost is extremely small in comparison to that obtained by eating food.

Utility. In the simulations of Chapter 6, actions and their outcomes are associated with a positive or negative utility. Typically, those utilities are fixed, being selected to reflect the real world to some approximation; however, in some experiments the utility functions themselves can evolve. The utility associated with an action can have a flow-on effect upon the agent's health. The health impact of utility is usually small relative to the direct health effect of an act; however, agents can occasionally derive a very large negative utility (as in experiments with rape). There are no examples of very large positive utilities, for the simple reason that an agent could use this as a substitute for eating.

4.5.3 Behavior

Agents are located in cells in the grid and often are oriented towards one of the eight cells in the Moore neighborhood. Actions, and the observations they may be conditioned upon, occur within this neighborhood or a larger $n \times n$ neighborhood, with the size dependent upon the experiment and the action or observation in question. Sometimes some actions are hard-wired, so that if an agent can perform them, it will. Usually, however, an agent's evolvable decision function selects an action, mapping *observations* of the environment to *actions*. Records of evolved decision functions are an important source of information about the adaptive nature of behavior in particular environments. We use several kinds of decision function through the course of our experiments, which we describe below.

Observations

What agents can observe varies with each experiment. In cases where observations are to be passed on to the evolvable decision function, this quantity is typically clipped and scaled (usually to the interval [0,1]) for purposes of uniformity. Observables include:

Self-health: the numerical value of the agent's health.

Self-age: the age of the agent in cycles.

Self-sex: whether the agent is male or female.

Local-population-density: the number of agents within the observing agent's $n \times n$ neighborhood, divided by the size of the neighborhood.

Global-population-density: population density taken over the whole world.

Local-food-density: food density within the neighborhood.

Global-food-density: food density taken over the whole world.

Mate-request: indicates whether there is another agent currently requesting a mate with the observing agent.

Is-gestating: indicates whether the agent is pregnant.

Food-availability: returns the value of the *fdf* divided by local-population-density.

Actions

Actions vary per experiment, however the basic set of actions is: reproducing, whether asexual, consensual or non-consensual; eating; movement; and resting. In addition, there are the acts particular to some experiments — namely, suicide, rape and abortion.

Reproduction. Simulations contain either asexual or sexual agents, and sometimes both. Where gender issues are unimportant, we sometimes use genderless sex (i.e., any two individuals within a species can mate). Often, however, we have males and females. The female sex is either the one that gestates or, when there is no gestation, the one that invests more in the offspring.[3] Whatever the form of reproduction, the offspring's initial health is taken from its parent or parents, as is its chromosome, after mutation. In sexual reproduction, there is sometimes a gestation period prior to birth; otherwise birth is immediate. Upon birth, the offspring is placed somewhere within the neighborhood of its parent, except when the user specifically

---

[3] In biology, the female of a species is the sex that has the larger gamete. Very frequently, the female is also the sex that invests the most overall.

CHAPTER 4. EVOLUTIONARY ARTIFICIAL LIFE 91

opts to turn off kin selection, in which case the offspring may be placed anywhere.

Mating in some simulations lacks the notion of consent, agents choosing partners, and then mating without asking. In other simulations, choosing mating entails making a request, which may be turned down. Acceptance, or non-consensual mating, will fail to produce offspring when the agent lacks sufficient health or maturity or when no viable partner is available, or sometimes when there is no available space for the offspring. In the absence of an alternative (i.e., consensual mating), non-consensual mating is not at all like rape, and there is no negative utility associated with it.

Eating. The agent eats a preferred or randomly selected piece of food in the neighborhood.

Rape. The agent forces mating with a victim, and the victim derives large negative utility.

Abortion. The female terminates its pregnancy prematurely.

Suicide. The agent causes itself to die immediately.

Movement. Agents can move (or, depending on the simulation, migrate or be transmitted) from one cell to another within their neighborhood. In some simulations, movement is a single action that takes the agent to a random neighboring cell. In others, it is divided into two actions: turning and walking. If, when trying to move, the agent is facing a hard boundary to the world or the cell ahead is full, the agent stays put.

Resting. When resting, the agent does nothing.

4.6 Evolution and the Agent Genotypes

Unlike evolutionary algorithms, our simulations have no explicit fitness functions and no neatly defined generations of agents. Agents are selected for by being better adapted to their environments than their competitors, and so are better able to reproduce. We may talk of generations in terms of the typical period of time agents require to mature and reproduce for the first time, as in real biology.

Agents over time may become better adapted to their environments. This is mainly reflected in improved decision functions encoded in their chromosomes, which here are either sets of **production rules** or decision trees. Both types have their own structure, usage, crossover method and mutation method. In addition to decision functions, agent genomes may also possess variables representing mutation rates, parental investments and an age of expiration. They may also hold bit signatures for an agent's immune system or mate compatibility and a disease's infectiousness or virulence. These elements are either described in the next sections or, in the succeeding chapters, in the descriptions of the experiments to which they pertain.

4.6.1 Production Rules

Structure and Usage. Production rules in those experiments consist of a fixed size set of rules, each rule consisting of a *condition* and *action distribution*. To make a decision, the agent tests the rules in random order to find a rule whose condition is satisfied by one of the available observations. The conditions are comparators (e.g., '>', '≤') and thresholds for the corresponding observations.

If the rule's condition is satisfied, the rule's action distribution is triggered. These are vectors of probabilities over the available actions. The probabilities are kept as non-negative floating point numbers, which are normalized when needed. Since these values are mutated, it is possible for them to go negative, which is interpreted as a zero probability during normalization. Once the action distribution is normalized, an action is selected by comparing a uniform variate from $U[0,1]$ with the distribution.

Crossover. When two agents reproduce, their decision functions are recombined by a multipoint crossover spanning both conditions and distributions (see Figure 4.1). The number of crossover points is itself part of the agent's genome, and so evolves. Conditions and individual action probabilities are treated as genes that are not broken by crossover. Alternating portions of the parents' rules, selected by the crossover points, are then included in the child genome.

Mutation. Mutations affect comparison values, action distributions and the number of crossover points. Mutations of floating point numbers are made by adding a random variate from a Gaussian distribution with a mean of 0 and a very small variance. If the result falls outside of an allowed range

CHAPTER 4. EVOLUTIONARY ARTIFICIAL LIFE

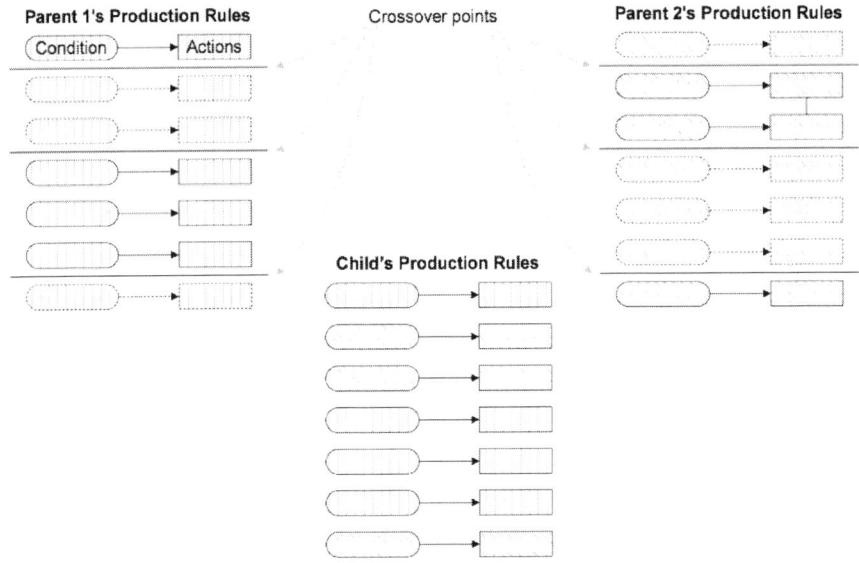

Figure 4.1: Crossover of fixed-length production rules.

it may be clipped.

4.6.2 Decision Tree

Structure and Usage. Decision trees consist of branches in which internal nodes describe conditions and leaves actions. In our simulations, the internal nodes describe observational conditions and the leaves are vectors of probabilities over available actions (i.e., action distributions). Thus, each branch amounts to a production rule in which the antecedents are conjunctions of observational conditions.

Crossover. Figure 4.2 shows an example of decision tree crossover. Crossover involves choosing one parent's tree as the main tree, and then swapping in a randomly chosen sub-tree from the other parent. The chosen sub-tree will keep its original location in the new offspring tree, or, if that location does not exist, it is placed at a random point.

Mutation. Mutations affect the action distributions, the branch types and the split values. Mutation of the action distributions occurs exactly as per their mutation in the production rules. Branch types are mutated according

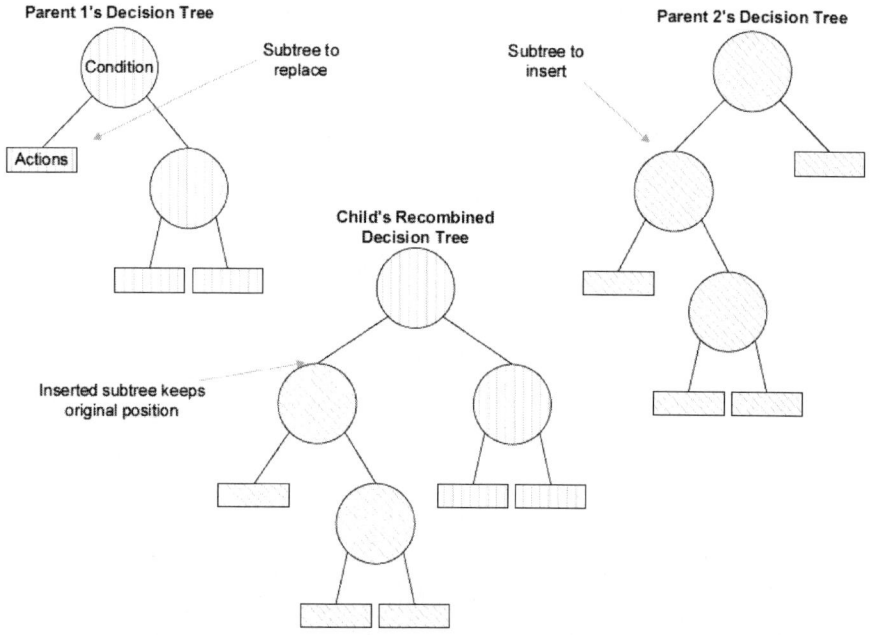

Figure 4.2: Crossover in decision trees.

to a Bernoulli distribution; if mutation occurs, another branch type (such as health) is uniformly randomly chosen to replace the existing type. The split value is mutated by adding a Gaussian variate to it.

Mutations can change the size and structure of decision trees. During crossover, a mutation probability may force the final location for splicing in a sub-tree to be higher or lower than its nominal target. Also, split nodes have a small probability of being removed (being replaced by a child) or added (e.g., with a leaf being changed to a split node).

4.6.3 Mutation and Meta-mutation

In general, there are mutation rates for each kind of thing that can appear in the genotype — e.g., comparators, comparison values, branch types, split values, action probabilities, crossover points, investment amounts and investment terms all have their own mutation rates. Wherever possible we encode these mutation rates within the agents' genomes; they are then inherited by offspring by random selection from one parent. Mutation rates are then themselves subject to evolution, driven by meta-mutation rates.

This has two major advantages: it eliminates multiple simulation dimensions that would otherwise have to be explored explicitly, and it allows the mutation rates to evolve to optimize the population's evolvability for the simulation environment, in analogy with biological evolution.

4.7 Statistics

There are several kinds of statistics that are generally useful across the simulations. These include demographics, action rates, and utilities. Statistics particular to each experiment are described in subsequent chapters.

4.7.1 Demographics

Generic demographical statistics kept include means and variances of age, population size and health, which help orient the interpretation of results.

4.7.2 Action Rates

An action rate is the per capita frequency with which agents perform an act. Sometimes the frequency and variation of action rates is the principle target of interest in a simulation, for example in the suicide simulations.

4.7.3 Total Utility

Total utility is used to evaluate the utilitarian value of an act in some environment. It is calculated by summing together the utilities that all agents derive throughout an epoch. To find out which of two simulations represents a more ethical world (in utilitarian terms), we can compare the *cumulative* total utility of the two simulations. This may be of interest when the two worlds differ *only* in the presence or absence of a particular kind of action, such as abortion.

4.8 Conclusion

The kind of simulation just described is the basis for our experimental work, with particular experiments incorporating particular variations from the basic architecture described above. Our experiments thus simulate a generic evolving population of animals. The animals eat, rest, move, procreate and engage in a variety of other activities, some of them fairly complex. They observe their environment and use these observations to condition their actions using on an evolvable decision function, which may be a set of production rules (variable or fixed in length) or a decision tree. These decision functions can be crossed-over and mutated — where the various mutation rates are kept with the agent, and subject to meta-mutation by the system. Finally, the system collects various demographic statistics, and statistics on action rates and on genetic properties.

This generic simulation design is both simple and extensible and can be adapted easily to new problems, as its many variations and uses documented in this book demonstrate. New actions, observations and decision functions can all be plugged in, new kinds of agents readily invented and other new entities aside from agents and food can be added, or the world topology changed. Indeed, we illustrate all of these options in our investigations of group selection and co-evolution in the next chapter.

4.8.1 Utilities in Agent-based Modeling

Agent-based modeling often incorporates decision-making. The essential property of a rational Bayesian decision-theoretic agent is that it assigns utilities to states of the world (including to states of itself) and chooses its actions for whatever will maximize its utilities. Such, at least, describes the rational *selfish* agent.

The utilities are normally fixed by the programmer, but they need not be. In principle, and in some of our simulations, utilities may evolve or be learned, adaptive in the literal or metaphorical sense, as we discuss in Chapter 5. In many of our simulations, utility is tied to decision making only implicitly, by side-effecting on the health of agents. In our experimental study of the evolution of utility, however, the utilities only role is the one envisioned in Bayesian rationality, of assisting the agent to maximize expected utility in its choice of action. While in our simulations of the evolution of utility we have, at best, evolved rational egoists, it would be interesting in the future to see if we can also evolve rational *utilitarians*. This would be in the most direct form the evolution of altruism, where the altruism is implemented not in the coin of specific behaviors rigged to be altruistic, but in the general capacity to understand the states of other agents and their value for those agents and responding by attempting to enhance them. Presumably, the conditions in which specific altruistic behaviors are evolvable can be recast in a more general representational setting so that altruistic decision-making capacity will evolve.

4.8.2 Simulating Ethics

The foregoing indicates how simple it is to extend an agent-based ALife or social simulation to encompass utilitarian analyses. All we need are utilities associated with the (dis)pleasures and (dis)satisfactions of agents and the rest is straightforward. While not possible in the real world, in the confines of a controlled simulation we can do our sums with the utilitarian equations of Chapter 2, allowing us to compare tallies for alternative worlds. The advantages of ethical investigation via simulation parallel those for biological investigation via simulation. In the terms of ALife, we can calculate utilitarian totals for silicon worlds as they are and real worlds *as they could be*. In silico, knotty concerns about the impropriety of experimental ethics vanish, while the only constraints on our ability to explore alternative histories are computing power and and our own creativity.[4]

ALife is a set of simulation *methods* more than it is a field of research. ALife's methods provide tools to investigate potentially any process in nature by creating a population of individuals, each with its own rules of behavior, and letting it run. Adding in methods of interaction between individuals, and especially reproduction with suitable mechanisms for copying, recombining and mutating their chromosomes, allows evolutionary ALife to be applied to investigating questions about evolutionary biology. Adding in also methods of cognition, decision making and value judgment by the individuals turns ALife into a powerful tool for exploring evolutionary psychology and the evolution of ethics. Many concepts from evolutionary psychology have already found their way into ALife simulations, helping them to produce behaviors that are more socially complex, interesting and informative. However, the most significant contribution is likely to be in the other direction, with ALife simulations helping us to understand and evaluate theories and ideas from evolutionary psychology and utilitarian ethics. That, at least, is the thesis we put forward in this book. We shall now attempt to support that thesis with what are, we hope, suggestive and supportive experiments in artificial evolution and ethics.

---

[4]To be sure, ethical concerns do not *entirely* vanish when working in silico. But the constraints are either really about limits upon freedom of speech or else they are thus far purely hypothetical concerns about the rights of artificial intelligences. Our agents, despite having some cognitive apparatus, are so far far too limited to raise this issue in any compelling way.

Chapter 5

Experiments in Evolution

We now apply the kind of artificial life simulation described in the last chapter to a number of issues fundamental for understanding the nature of evolution. Before looking at any specific experiment, we consider the particular, and central, question of the level at which selection pressure occurs. Many levels have been proposed: the gene, the cell, the organism, the group, the species. The traditional level of selection has been located at the individual, and this has been vigorously defended as the exclusive locus of selection pressure for most of the history of evolution theory. It has lost its grip in recent times, due in part to the multilevel selection theory of David Wilson, which represents the many levels as a kind of Russian doll. Another factor has been a growing number of successful computer models of group selection, such as those we present after the next section.

5.1 Levels of Selection

Individual selection refers to selection pressure in favor of the organism with the greatest individual fitness, meaning the ability to survive and reproduce, or more simply and directly the individual's expected number of descendants. Naturally, early aging — that is, an early onset of senescence and death — is generally maladaptive, implying carriers of the trait are relatively individually unfit. Still less adaptive, indeed dramatically maladaptive, is any propensity to commit suicide. For much of its existence evolutionary biology has focused upon individual selection as the only, or perhaps primary, form of selection pressure, and so the evolution of aging, suicide, or more generally altruism were anomalous. And yet all three are demonstrable in nature and all are at least arguably adaptations in at least some of their forms. Aging rates in animal and plant species vary greatly and systematically, ranging from a few hours for some phytoplankton cells (Agustí et al., 1998) to a few days for some insects to thousands of years for the bristlecone pine tree. The variations in life span across species is hugely greater than the variations within species; it seems almost certain that selection pressure of some kind is keeping the aging rates of particular species within fairly narrow bands. In other words, in cases where the maximum life span has evolved downwards, or been held down, presumably directly damaging individual fitness within the population, there is likely to be an adaptive reason for it.

The adaptive benefits of early aging or suicide, if there are any, could only be received by organisms other than the organism exhibiting the trait. For this reason, such a trait, if beneficial and adaptive, must be altruistic. In order to give an explanation of such an altruistic adaptation, one must call upon a mechanism of selection which incorporates such altruistic benefits, or otherwise deny that it is an adaptation.

There have been two notable attempts at explaining such a mechanism, namely group selection, generally attributed to Wynne-Edwards (1962), and inclusive fitness theory, proposed by Hamilton (1964). Group selection differs from individual selection in that it is the group rather than the individual organism that selection acts upon. *Groups* are variously defined in the literature; we shall generally mean actively interbreeding subpopulations, which in biology are called **demes**. Inclusive fitness (or kin selection) theory, by contrast, shifts the focus downwards from the individual to the gene (or allele), whether held by the individual or, as a replica, by a relative. The inclusive fitness of a gene is just the organism's individual fitness augmented by the harms and benefits caused to the fitness of others, weighted by their relatedness, i.e., the probability of their carrying the same allele due to common ancestry (Hamilton, 1964). This gene selection mechanism is often labelled kin selection, since the probability of two individuals carrying the same allele due to common ancestry is a function of their kinship (see Figure 1.2).

Inclusive fitness theory has become widely accepted, especially as an explanation for the evolution of altruistic behavior. The group selection concept, however, has remained more contentious. It has been doubted, for example, whether the selection pressure for selfish behavior within groups can be overcome in nature by selection pressure for altruism between groups. In the succeeding sections we describe both of these mechanisms. We then consider the relationship between them, often held to be one of opposition.

We see them, however, to be in a supportive relationship of a kind that recurs repeatedly in artificial life simulation, namely one of supervenience.

Group Selection

The group selection concept has an enduringly bad reputation for many. Introduced by Darwin (1880), it was widely used by evolutionary biologists until the highly successful critique of Williams (1966), which raised doubts that a workable model of group selection could be devised. While perfectly sensible models for it have been developed in the meantime, it still carries the taint from Williams' attack.

Early Group Selection Models. The first workable group selection model was proposed by Gilpin (1975) and ascribed group selection a major role in the evolution of population regulation. In this model it is the differing viability of the groups, together with their fecundity, i.e., their tendency to establish new groups, that drives group selection. Selfish individuals within groups may be more fit within the group, however the group as a whole pays a price in the degradation of its average fitness; that is, the group survives a shorter time than more altruistic groups and so establishes fewer colonies. Selfish individual fitness flourishes, but only within the group and its few offspring; even though altruistic behavior may be less fit within the group, it can be more fit across the population if groups favoring it are themselves more fit relative to other groups. This kind of group selection model was reviewed by Maynard Smith (1976) using a simplified version of Gilpin's (1975) predator-prey model (Figure 5.1).

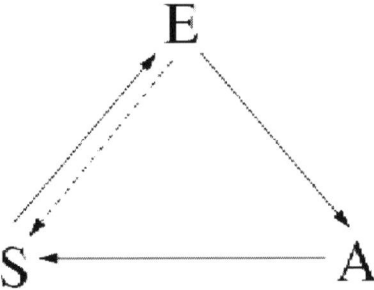

Figure 5.1: States and transitions of the early group selection models.

Maynard Smith's model contains some number of discrete patches, each capable of supporting a single group. There are three different states each patch can take at a given time: empty (E), containing no group population; selfish (S), holding a group that contains *at least some* selfish individuals; or altruistic (A), holding a group that contains *only* altruistic individuals. Transitions occur between states due to extinction ($S \rightarrow E$), migration or re-population ($E \rightarrow S|A$), or the introduction of selfish genes by mutation ($A \rightarrow S$).[1] In these models the main question is how the cooperation of the altruistic individuals will affect the factors of extinction and migration.

Maynard Smith (1976) observes that the fate of these models, when viewed in this manner, is dependent on the single parameter M, which is "the expected number of successful 'selfish' emigrants from an S patch during the lifetime of the patch." A successful selfish emigrant is one that establishes itself and leaves descendants in a neighboring E or A patch. If the average number of emigrants from S patches is greater than one ($M > 1$) then the S patches will increase in frequency. Otherwise, if $M < 1$, the S patches will become extinct faster than they can found new groups and will therefore be selected out of the system. (Of course, the ratio of selfish to altruistic genes in the total population also depends upon the expected number of colonizations by A patches.) This model demonstrates a mechanism of group selection that is clearly a logical possibility. However, it has been loudly doubted whether the supposedly stringent conditions required for this kind of evolution of altruism could be realized in nature by, for example, Maynard Smith (1976) himself.

Species Selection. Species are themselves reproductive groups, and there is a real possibility that altruism may have evolved by group selection in the form of **species selection**. The concept of species as a taxonomic classification is central to biology. Unfortunately, the species concept, just as the even more fundamental concept of life, resists efforts to capture it in any precise definition (Hey, 2006). There seems to be an ineradicable vagueness to it, exhibited in application to, for example, asexual species, ring species and hybrids. The most generally accepted definition of species, which we follow here, is that of a reproductively isolated subpopulation (Mayr, 1963) — that is, a group of actually or potentially interbreeding organisms that are reproductively isolated from other such groups. The fact that this manifestly fails to deal with asexually reproducing species will not concern us here: this definition suffices for most of our evolutionary simulations. Per-

---

[1] The transition $S \rightarrow A$ by mutation can be ignored, since it requires simultaneous mutations throughout the group.

CHAPTER 5. EXPERIMENTS IN EVOLUTION

haps it is worth noting that this definition does not exclude homosexuals, or even the reproductively handicapped, from being counted within the species in which they are born; the *potentiality* of reproduction can be interpreted liberally and counterfactually.

The geographic situations in which speciation can occur have given rise to its terminology: **allopatric** and **peripatric speciation** rely on the geographic isolation of subpopulations, whereas **sympatric** and **parapatric speciation** are based on the emergence of new species with little, or no, geographic isolation, due, for example, to behavioral divergences between subpopulations that renders them reproductively isolated.

Eldredge and Gould (1972, p. 84) introduced the idea that evolution proceeds through long periods of genetic stasis, punctuated by short periods of rapid change, or punctuated equilibrium (PE) theory:

> The history of life is more adequately represented by a picture of "punctuated equilibria" than by the notion of phyletic gradualism. The history of evolution is not one of stately unfolding, but a story of homeostatic equilibria, disturbed only "rarely" (i.e., rather often in the fullness of time) by rapid and episodic events of speciation.

Eldredge and Gould contrasted punctuated equilibrium with what they saw as the established but mistaken view that evolution can only occur gradually or even only at a constant, continuous rate — **phyletic gradualism**. While acknowledging that the fossil record cannot definitively settle the question of whether most evolutionary change occurs in punctuation events, they asserted that PE theory better matches the paleontological record. Whereas the fossil record is limited in its ability to support or undermine PE, our artificial life simulations can provide clear tests of whether or not PE is actually occurring within different simulations.

Under the punctuated equilibrium concept, once a species becomes static and defined, it takes on a kind of individuality. It has a lifespan; it has the opportunity to reproduce through further speciation; and, in the end, it will die. This supports a strong analogy with individual reproduction and, therefore also, with individual fitness (Gould, 2002). But the similarity is more than analogical with group selection, for this just *is* a kind of group selection. Species become units of selection, competing with other species within the biosphere for the opportunity to create new species and to avoid early extinction; this creates a species selection mechanism which falls under the group selection model described above, and which caters for the evolution of altruistic traits.

Kin Selection

Inclusive Fitness. Shortly after Wynne-Edwards proposed his group selection model Hamilton (1964) developed inclusive fitness theory, which was initially seen as an alternative method of explaining the evolution of altruism. Inclusive fitness theory shifts the selection emphasis from the individual to the allele, whether within the individual or as a replica in another. Fitness benefits accruing to any copy are reflected in its frequency within the reproducing population. Since copies are more likely to be found within kin, the evolution of altruism via kin selection is also a real possibility. While this suggests that behavior specifically focused upon kin may be favored, identifying or explicitly responding to kinship is not necessary for differential kin selection pressure to arise. The benefits of altruistic acts quite commonly favor those in the vicinity of the actor, and these are usually more closely related to the actor than those in no position to benefit.

Hamilton's inclusive fitness rule states that the criterion for the positive selection of an allele is:

$$\sum_i (b_i - c_i) r_i > 0 \qquad (5.1)$$

where the subscript i denotes the ith member of the species, r_i is the relatedness coefficient between actor and individual i, b_i is the benefit to the fitness of the individual i, and c_i is the cost to the fitness of the individual i. The relatedness r is proportional to the chance that a copy of the same allele at a given **locus** will be held by both the donor and recipient. For example, siblings have an equal chance of inheriting the alleles of either parent at a particular locus and hence have a relatedness of $\frac{1}{2}$. Considering the simplified case of an allele which only bestows benefit on a single sibling s at some cost to its carrier a, applying (5.1) we can see that for an allele to be selected for $b_s \times \frac{1}{2} - c_a > 0$. That is, the benefit to the receiving sibling must be greater than twice the cost to the donor. The consequences of the theory are summed up by Hamilton (1964) in two points:

1. For an allele to be selected for it is not sufficient that it should increase the fitness of its bearer above the average if this is done at the expense of related individuals. Being selfish does not imply being fit.

2. Conversely, an allele may be selected for while disadvantageous to its bearers, if it sufficiently benefits relatives.

The Supervenience of Group Selection on Kin Selection. Inclusive fitness theory was considered an incompatible alternative to group selection for explaining the evolution of altruistic behavior until Price (1970) and Hamilton (1975) reformulated it into equations which analysed the different levels of selection, namely within- and between-group selection, into two distinct terms which sum to explain total change in the frequency of an allele.

Hamilton defined his groups as sharing *equally* the benefits of altruism, creating an important difference between his idealized groups and real groups. As there is no within-group preference for holders of the same alleles, such as kin, altruistic alleles are necessarily selected *against* within a Hamiltonian group, as any selfish free riders receive the same benefits as everybody else without paying the cost. This assumption is also made implicitly in the group selection models discussed earlier. In these models, for an altruistic allele to be positively selected, the magnitude of between-group selection must be greater than within-group selection. That is, groups with a higher frequency of altruists can perform better by increasing the group's fitness and hence increasing the size and proportion of altruistic groups in the population. If this increase outweighs the decrease in frequency of altruists within each group, altruism will increase in global frequency. The more varied the frequency of altruists across the groups, and the more benefit bestowed by the altruists on the group, the greater this between-group selection pressure will be. Figure 5.2 illustrates this effect in a population divided, for a period, into two groups with a varied frequency of altruists, represented by slices of pies. Within-group selection causes the altruistic "pie slice" to shrink in both groups. Whereas, between-group selection causes an increase in the altruistic "pie size" resulting in an increase in the overall frequency of altruist genes.

The relationship between the **Price equation** and group selection is considered by some as simply one of mathematical convenience (e.g., Maynard Smith, 1976). Wilson (1980), on the other hand, uses the equation as the basis of his multilevel selection theory, which asserts the compatibility of multiple levels of selection that are simultaneously active. Hamilton (1975) argued that, due to complex kin associations, with isolated groups relatedness eventually builds up to:

$$r_g = \frac{1}{2E+1} \quad (5.2)$$

independently of group size, where r_g is the mean intra-group relatedness and E is the absolute number of emigrants per group, per generation. That is, virtually closed groups become highly related units, regardless of size.

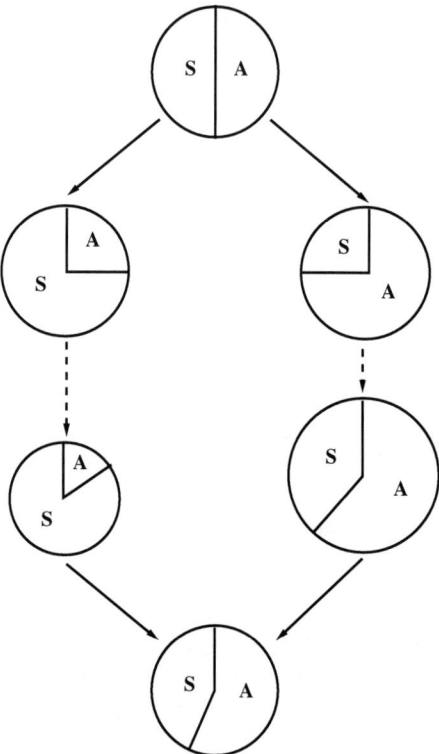

Figure 5.2: Within- and between-group selection (adapted from Sober and Wilson, 1998).

On the controversy over the relative importance of group selection versus kin selection for the evolution of altruism Hamilton (1975) remarked:

> Because of the way that it was first explained, the approach using inclusive fitness has often been identified with 'kin selection' and presented as an alternative to 'group selection' as a way of establishing altruistic social behavior by natural selection... Kinship should be considered just one way of getting positive regression of genotype in the recipient, and that it is positive regression that is vitally necessary for altruism.

The idea is that inclusive fitness is more general than kin selection and might arise by mechanisms other than kin selection. What is needed for group selection of altruistic behavior is, first, a variable distribution of altruism between groups and, second, a positive association between group fitness and the altruistic genes, as required by the Price Equation. Such an association can arise by kin selection or by other means, but, however it arises, it will result in differential group fitness leading to a spread of altruism.

Kin selection is one process that may lead towards establishing altruistic behavior within a population of multiple groups. In particular, it can prevent selfish behavior from simply dominating all groups, precisely because the Hamiltonian idealization of the benefits of behaviors being accrued equally by all group members is quite generally false. Furthermore, kin selection is unlikely to act in all groups with equal force and rapidity, so the requisite variability in altruism across groups arises naturally. With this basis, group selection forces can operate upon altruistic behaviors that are *already* actively selected for by kin selection. In Wilson's multilevel selection theory, the different levels of selection operate independently of each other (Wilson, 1997). What we are here describing is a relation of supervenience, where group selection supervenes upon kin selection (while, of course, requiring other ingredients as well, such as the presence of multiple, semi-independent, interbreeding subpopulations). As Hamilton noted, other mechanisms leading to the conditions needed for group selection are also possible, but kin selection is indeed one of those mechanisms. Furthermore, it seems to be the most natural one. The only clear alternatives are those of associations between variable altruism and group longevity arising by chance and the artificial one of intervention by a designer. The former requires a reasonable number of stable, small groups (since with large groups the law of large numbers will eliminate the required variation); the latter requires a virtual simulation environment or, in any case, a designer who likes to play with evolution. In fact, in many of our simulations in which group selection played a role, when we turned off kin selection (see next paragraph), group selection — *all* selection pressure — disappeared with it. It is plausible that in many situations, although kin selection remains logically unnecessary for group selection, it may be practically necessary: as in many of our simulations, it may well be that nature provides no alternative basis for differential group survival and reproduction. Furthermore, the active operation of kin selection in establishing some of the conditions needed for group selection surely undermines the common skepticism about nature providing those conditions. Thus, Hamilton's (1975) suggestion that we reserve the term 'group selection' for situations where kin selection is inoperative would empty the term of practical application.

The Kin Selection Button. In many of our simulations, including those accompanying this book, we have a "kin selection button" for turning kin selection on or off. This allows users to see directly the effect of inclusive fitness (when kin selection is active) and contrast it with the effect of non-inclusive fitness (when kin selection is turned off). Typically, the difference is dramatic, because the simulations we are interested in frequently have altruistic behavior at their center, leaving individual fitness unable to cope.

The way in which kin selection operates in our simulations is via the correlation between proximity and kin relatedness. That is, so long as agents do not have very high rates of motion, which would tend to spread them over the world randomly, two agents who are near each other are more likely to be closely related than two agents at distance. So, actions which are selectively beneficial to neighbors will also be selectively beneficial to relatives. As Maynard Smith (1976) noted, "kin selection can operate whenever relatives live close to one another, and hence can influence one another's chances of survival and reproduction" — which, in fact, is the normal state of nature. Thus, one way of turning *off* kin selection is to redistribute agents randomly around the world.[2] Alternatively, one can simply spawn agents at random locations around the world. Unfortunately, either such direct approach to eliminating kin selection would also radically alter the population dynamics of the simulation. For example, in the case of a Gilpin-type model (Figure 5.1) randomly spawning children to empty patches would negate the effects of migration founding new groups. To address this problem we adopt the mechanism of a compulsory adoption queue. As new offspring are born to a set of parents they are placed at the end of the adoption queue and replaced by one from the front of the queue. Thus, the simulator can switch kin selection from "on" to "off" and still ensure that all other factors, including group structures and group sizes, remain unaffected.

5.2 The Evolution of Aging

Aging is defined as the general deterioration of an organism, and its eventual death, by internal causes (Williams, 1957). The rates at which different species age is a perplexing phenomenon. As we noted above, the divergence of life spans across species is extraordinary, and these different aging rates have themselves not varied significantly during recorded history,

---

[2]This might be insufficient for agents who are capable of *recognizing* their kin and acting accordingly. Our agents have yet to evolve such advanced cognitive capacity, however.

CHAPTER 5. EXPERIMENTS IN EVOLUTION

so far as we can tell. The genetic control of aging is beginning to come into view, with multiple genes already identified as participating in aging rates (e.g., Belenky et al., 2007). All of this seems to suggest that aging rates have evolved because of their adaptive value. However, the obvious fitness costs of fast aging on individuals would cause strong direct selection pressure against it, suggesting that aging may be a side effect of some more essential characteristic, i.e., that it is non-adaptive. Historically, both adaptive, e.g., Weismann (1889), and non-adaptive, e.g., Williams (1957) and Medawar (1952), explanations of aging have been proposed. Recently there has accumulated compelling experimental evidence that aging *is* an adaptation (Mitteldorf, 2004, Bredesen, 2004, Skulachev, 1997). This has lead to a resurgence of research into possible adaptive benefits of aging, including our own.

Adaptive Theories of Aging

Weismann (1889) was one of the first biologists to publish an evolutionary explanation of aging, holding that aging is an adaptation, i.e., selected for its own sake. In recognition of the obvious direct cost of aging to the individual, Weismann argued that aging death is beneficial to the species as it removes the worn out individuals which "are not only valueless to the species, but they are even harmful, for they take the place of those which are sound" (Weismann, 1889). Weismann believed that over time an individual would be unable to avoid accumulating slight injuries, causing it to become defective and crippled, and affecting its value to the species. This is contrary to the claims of some who say that Weismann claimed the benefit of aging was the adaptability of the species to a changing environment (Goldsmith, 2004, Mitteldorf, 2006). While this is our own view of the benefit of aging, supported by our experiments, we have seen no evidence that Weismann shared it.

Williams (1957), responding to Weismann's hypothesis, summarized the most forceful of the criticisms of the theory:

1. the fallacy of identifying senescence with mechanical wear;

2. the extreme rarity, in natural populations, of individuals that would be old enough to die of the postulated death mechanism;

3. the failure of gerontological research to uncover any death mechanism;

4. and the difficulty of conceptualizing how such a feature could be produced by natural selection.

In the first criticism, Williams argues that Weismann's proposed benefit of aging hinges on the assumption that the older members of the species will have deteriorated, regardless of senescence. He contends that Weismann's argument is circular, that is, it assumes what it claims to explain, which is declining vigor with age. The remaining criticisms outline potential difficulties with any adaptive theory of aging, without, however, falsifying the thesis.

Overall, Weismann's hypothesis was heavily criticized, particularly for being unable to translate its proposed fitness benefits of aging into individual fitness. Consequently, critics have held that aging has an inherent negative effect on individual fitness and therefore must be non-adaptive. Clearly, after the demonstrated mechanisms of kin and group selection, there is an opportunity to reconsider the historical assessment.

Non-Adaptive Theories of Aging

The theories of antagonistic pleiotropy and mutation accumulation are among the more popular non-adaptive evolutionary explanations of aging. Antagonistic pleiotropy, proposed by Williams (1957), holds that aging is a side effect of genes which are selected for other beneficial properties. He assumed that (1) there exist pleiotropic genes which have opposite effects on fitness at different stages of the bearer's life-cycle and (2) that the probability of reproduction decreases with age, which follows from external causes of death and injury in the natural environment. He argued that pleiotropic genes which display benefits earlier in life could be selected for regardless of self-destructive costs associated with the gene later in life.

Mutation accumulation, proposed by Peter Medawar (1952), holds that aging is the result of mutational load. Medawar argued, based on **Fisher's reproductive value** (Fisher, 1930), that aging was the result of random, detrimental mutations which show their effects only late in life. Reproductive value is the expected future reproduction of an individual of a certain age — a function of the probability of survival and rate of reproduction. Fisher stated that the pressure of natural selection must be proportional to reproductive value, while an organism's reproductive value decreases with age, leading to a diminution of selection pressure. Medawar concluded from this that deleterious alleles that affect the organism later in life will arise by mutation faster than they can be selectively removed from the population, because selection pressure against deleterious alleles manifesting in later life is weak. In this, Medawar's theory is similar to antagonistic pleiotropy theory, that is, by assuming decreasing selection pressure with age.

Skulachev (1997), Mitteldorf (2004) and Bredesen (2004) argue that, although these theories had good support forty years ago, there has since accumulated experimental evidence which is inconsistent with both. For example, pleiotropic theories, which imply a direct causal link between fertility and aging, are undermined by experiments that increase longevity without any apparent cost to fertility (Leroi et al., 1994). Also, caloric restriction experiments demonstrate that animals can forestall aging, without any fitness cost, in times of dietary stress (Weindruch and Walford, 1986), which raises the question why aging is not forestalled all the time. This has lead to a resurgence of research into the possible adaptive benefits of aging, including Mitteldorf's demographic theory (2006), Skulachev's phenoptosis theory (1997) and our own theory, which we discuss next.

The Diversity Hypothesis

Our hypothesis, as that of Weismann, postulates an adaptive explanation of aging, that it is selected for its own fitness benefit. We suggest that aging has a group fitness benefit which can outweigh the individual fitness costs. Groups with shorter individual life spans will turn over faster, introducing variation through recombination and mutation faster, and consequently have greater diversity. This will confer benefits to the groups with faster aging in situations where new adaptations are needed, such as in periods of rapid climate change or co-evolutionary competitions. In co-evolution scenarios, e.g., predator-prey and host-parasite interactions, groups with greater diversity will be less easily exploited, creating a stronger and healthier subpopulation.

This hypothesis is related to Matt Ridley's Red Queen Hypothesis (Ridley, 1993), explaining the evolution of sexual reproduction. On the face of it, asexual reproduction should lead to greater fitness compared to sexual reproduction. Sexual reproduction is far more complex to negotiate: in addition to requiring more complex reproductive equipment (such as the production of gametes), it requires organisms to locate a suitable mate and carry out a more complex mating act. However, as Ridley emphasized, species engaged in co-evolutionary races with predators and parasites require continued infusions of new and diverse genetic material for selection to operate upon, just to keep up in the co-evolutionary race ("It takes all the running you can do, to keep in the same place," said the Queen). Ridley's hypothesis suggests that the diversity benefit of sexual reproduction, in allowing the recombination of two parents' genes in addition to copying error mutations, outweighs even the substantial obstacles to carrying it out successfully.

In the following sections we discuss mechanisms of selection which potentially favor altruistic traits and use them in the design of an ALife simulation of host-parasite dynamics. Guided by Williams' criticisms of the Weismann hypothesis (see above), we conduct experiments with our simulation, demonstrating the evolution of aging and validity of our hypothesis in those simulation environments.

5.2.1 Comparing Alternative Aging Hypotheses

Simulation Design

In order to test explanatory hypotheses for aging we developed a multi-agent ALife simulation environment of the type described in Chapter 4, but tailored to our problem here, in particular by supporting multiple groups on separate patches. In our design description here we will emphasize the special features of the aging simulations.

Agents interact within groups sharing a food source and reproducing, potentially sexually. Groups may also interact with each other by mechanisms of group extinction, via disease epidemics, and migration (pioneering new groups), catering for a group selection mechanism (see §5.1). The simulations testing our diversity hypothesis have groups containing co-evolving populations of host and disease agents. Host agents interact with each other, through mating, and with the disease agents, which parasitize them. The success of disease transmission between hosts is determined by the genetic similarity of the host chromosomes.

When designing simulations it is necessary to consider the trade-off between the complexity of the model and its completeness. More complex models are harder to analyse; however, simpler models may neglect important mechanisms which allow validation against real systems (Grimm et al., 2005). In designing our aging simulation we have attempted to provide enough detail to support meaningful tests of the various contending hypotheses and not more.

Aging World. The simulation consists of a 10×10 grid of patches, with the edges connected forming a torus-shaped world. Each of the group locations may contain a group of hosts. The occurrence of group extinction and

CHAPTER 5. EXPERIMENTS IN EVOLUTION

migration are controlled by within-group factors, primarily through their population size. When an extinction occurs, the patch contents are reset and it becomes available for a pioneering population. When a group produces a migration party, two agents are randomly selected from the population and a destination patch is randomly selected from the immediate 3×3 Moore neighborhood. In the interests of simplicity, agents are not permitted to migrate to occupied locations, "invading" existing groups. Instead the destination group is selected from the unoccupied neighboring groups, unless there are none, when the migration attempt fails.

Groups. Each occupied patch contains a food store and a host and disease agent population. Energy within groups, as in real world ecosystems, is closed — energy can enter the group via food production (i.e., ultimately from sunlight) and only exits via agent overheads (e.g., heat). As a simple and effective technique for managing simulation resources, we impose the rule of conserving energy in the world. When an agent dies, this is performed locally, with any energy it holds recycled *indirectly* through the patch's food store, via food fertilization.[3] This is especially important for our aging simulations, since otherwise we would be actively penalizing groups with higher agent turnover. Each cycle the host population feeds directly from the food store, and the disease population, if one exists, feeds on the host population.

Host Population. The host population is the focus of the simulation, and it is this population which determines when extinction and migration occurs. Group extinction occurs when the host population completely dies out. Usually this will be caused by a disease. At the beginning of each cycle every group is tested, in random order, for the production of a migration party, determined by a probability proportional to the size of the host population, such as $pr = population\ size \times 10^{-3}$. Host populations which have greater resistance to disease transmission and infection will be greater in size, as less energy is diverted into sustaining the disease population.

Agents compete in a single location, on a purely individual selection basis, with patches having no internal structure. A kin selection effect could have been catered for by providing a spatial structure within the groups,

---

[3] A strange side-effect of doing this kind of research is being the recipient of outlandish accusations. One such was that we were simulating *cannibalism* because of our enforcing conservation of energy. That is much like accusing earth's ecosystem of cannibalism! If we wanted to simulate cannibalism, we would do so without any kind of ambiguity.

with kin more likely to affect each other, but this was considered unessential for this experiment and was omitted for simplicity.

Host Agents. Each host carries a chromosome inherited at birth, which is used to determine the genetic characteristics of the agents, covered in the following discussion. The age, health and fertility of the hosts are updated each cycle and are used, in conjunction with relevant genetic characteristics, to determine the behavior of the carrier, in any given cycle. There are four different events which host agents may choose or undergo in any cycle: migrating, eating, mating, and dying. Migration is tested for every group at the end of the cycle, after all agents have been updated for other events.

Host Eating. Each cycle every host is allocated an equal share of the group's food store. This food is used to replenish the host's health and fertility stores. Health represents the agent's energy, i.e., while positive the agent is still alive, and fertility is used for parental investments and to test maturity. The host's health, initialized at 10 units at birth, is decreased by a one-unit overhead every cycle. Usually food energy is diverted directly into the health store, however, in antagonistic pleiotropy experiments, the input energy is divided between health and fertility stores as determined by a gene. When either of the stores exceeds its capacity of 10 units, excess food is first diverted to the other store and then, if any food remains, it is recycled back to the group food store.

Host Reproduction. The hosts reproduce sexually but are genderless. An agent that chooses to mate is first checked for maturity, i.e., if its fertility store exceeds the threshold of five units required for parental investment. If the host agent is mature, it randomly selects a mature partner from its group and they then produce an offspring, with an inherited chromosome and health initialized at 10 units, half of which is taken from each parent's fertility store. The child is then added to the group population and acts independently in the next cycle.

Host Death. At the end of the cycle, each host is tested for death. The agent dies:

1. in case of accident;

2. when it exceeds its genetic expiry age;

CHAPTER 5. EXPERIMENTS IN EVOLUTION 115

3. in the event of an age-triggered death mutation;

4. or if it finishes the cycle with zero or negative health.

The genetic age of expiration and death mutation both cause age-related deaths. The former is caused by an aging-rate gene, while the latter is a mutation-accumulation mechanism. In order to test claims related to accidental deaths, we implemented deadly accidents which can occur at any time during the agent's life span, at a fixed probability of 0.1 at each cycle. We placed an overall finite bound on ages of 200 cycles, which, due to the high rate of accidental death, had a negligible effect on death rate. Health deaths can occur due to overcrowding, disease epidemics or failure to maintain the health store (which only occurs when antagonistic pleiotropy genes are enabled). When the agent dies, it is removed from the population and its health and fertility are recycled.

Host Chromosome. The host chromosome contains: a genetic expiry age; an antagonistic pleiotropy gene; a mutation accumulation string; and a vulnerability (immunity) string. Components may be evolving, fixed or inactive depending on the experiment conducted.

Genetic Expiry Age Component. In all experiments the expiry age gene, which is the focus of the study, is always active, and either fixed or evolving. This gene is used at conception to determine an expiry age for the agent by sampling a normal distribution with variance proportional to its magnitude, i.e., $N(expiry\ age, (\frac{expiry\ age}{3})^2)$. The gene is inherited from one of the parents, selected randomly, and mutated by a normal distribution with variance proportional to its magnitude, i.e., $N(expiry\ age, (\frac{expiry\ age}{100})^2)$. As this gene has no side effects and is not linked during reproduction to any other gene, fast aging must always be selected against unless the scenario provides aging its own selective value.

Antagonistic Pleiotropy Component. When antagonistic pleiotropy is enabled, a gene is used to determine the ratio of food energy directed toward health (maintenance) and fertility. This gene is inherited at random from one of the parents and mutated by the normal distribution $N(0, 0.01^2)$. In most simulations, the rate of health death is negligible, due to the high rate of accidental death. However, we expect a synergy effect with disease epidemics as a population with the pleiotropic genes will be, on the whole, less healthy and therefore more prone to epidemic extinction.

Mutation Accumulation Component. We simulate mutation accumulation with chromosomes containing a string of "death" bits. Each cycle, this death gene is sampled at the index of the agent's age, and, if on, death occurs with a 0.5 probability. The string length is 200 bits, covering all possible ages. The death string is inherited by crossover of the strings of both parents, with a 0.001 probability of mutation at each locus. This gene's role is similar to that of the genetic age of expiration. However, unlike the expiry age gene, the selection pressure at each position is dependent upon the probability of reproduction decreasing with age (see Section 5.2), which certainly obtains in our simulations due to accidental deaths.

Disease Agent. For testing our diversity hypothesis we introduced disease agents and host vulnerability strings, adapted from Epstein and Axtell (1996). The disease agents parasitize the hosts, using host health for the production of their offspring. The disease agents consist only of a chromosome containing an infection bit string, which determines whether the disease successfully infects a host given its vulnerability bit string (discussed below). The disease, which only lives for one cycle, does not store energy, but instead uses it immediately to cover overheads and to reproduce.

The Disease Lifecycle. The disease agents are updated each cycle asexually, producing offspring which are transmitted to neighboring hosts within its group. The number of disease offspring is determined by a split rate, with a default probability 0.15 of generating two, and otherwise generating one. The disease offspring immediately absorbs its energy requirements from the new host's health store, decreasing it by 1 unit. If the host already has non-positive health, the disease transmission fails. Further infection is then determined, discussed in detail below, and, if successful, the offspring disease is saved for the next cycle. At the end of the cycle the parent disease is overcome by its host and dies.

As disease agents are not transmitted between groups, a new disease is generated for each group at the beginning of every cycle to ensure that disease populations are never completely eradicated. (Note that we are not studying disease dynamics here, but instead focused upon the evolutionary response to disease in hosts.)

Disease Infection. The success of infection is determined by interactions between the host vulnerability (lock) and disease infection (key) bit strings (see Figure 5.3). The chromosomal host vulnerability string is 50 bits long and is determined by uniform crossover of its parents' vulnerability strings, with, usually, a 0.01 probability of mutation per bit. The disease infection bit string is five bits in length and is derived by copying the parent's string, with a 0.01 probability of mutation per bit. An infection attempt is successful when the infection string matches any contiguous sub-string of the host's vulnerability string. Hence, the more genetically similar the host agents are within a group, the greater the chance of successful transmission. The host vulnerability mutator determines the diversification rate, i.e., the diversity introduced by each new generation.

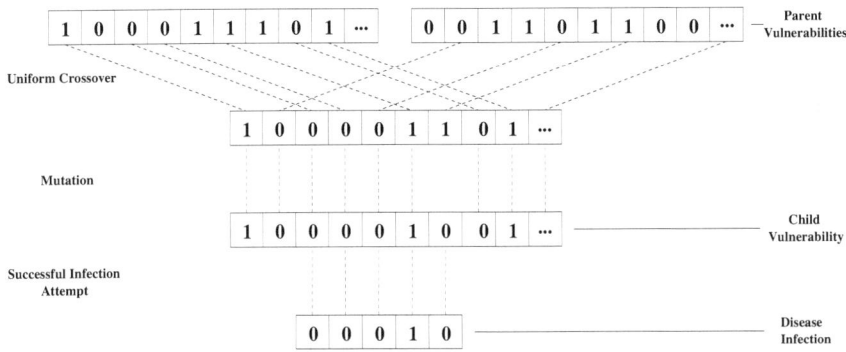

Figure 5.3: The child's vulnerability bit string is determined via uniform crossover of the parent's vulnerability strings, with a chance of mutation flipping each bit value. An infection attempt is successful when the disease's infection bit string matches a substring of the host's vulnerability string.

Aging Experiments

In investigating our diversity hypothesis for aging we first examined the within-group dynamics of the host and disease populations and the ability of the host populations with fixed aging to resist exploitation by the disease population. Next, we conducted a series of simulations with differing combinations of group selection pressure, diseases, antagonistic pleiotropy and mutation accumulation. These demonstrated a variety of possible worlds supporting the evolution of distinct aging rates.

Experiment 1. Initially, experiments were conducted to measure the effects of varied genetic expiry age (5, 20, 50) and vulnerability mutation rate (0.000, 0.005, 0.010, 0.015) on group diversity or "relatedness". This was done by conducting simulation runs with the disease component disabled, measuring the group relatedness at the end of 5,000 cycles (see Figure 5.4). Twenty simulations with different random seeds were run per parameter set in order to measure the variability of results. The group relatedness, the inverse of diversity, was measured as the average frequency of vulnerability string locations which were the same (i.e., the inverse Hamming distance) between two randomly selected hosts within a group. Thus, a relatedness of 0.5 means a perfectly mixed group, whereas 1.0 is genetic homogeneity. Simulations were not conducted with very low fixed expiry ages, giving host agents at least a short amount of time to reproduce themselves. Figure 5.4 shows that groups with lower genetic expiry ages (i.e., greater population turnover) and a greater vulnerability mutation rate had, on average, a lower group relatedness, as expected. (Unless specifically noted to be otherwise, any statistical results we comment upon were found to be statistically significant at the 0.05 level.)

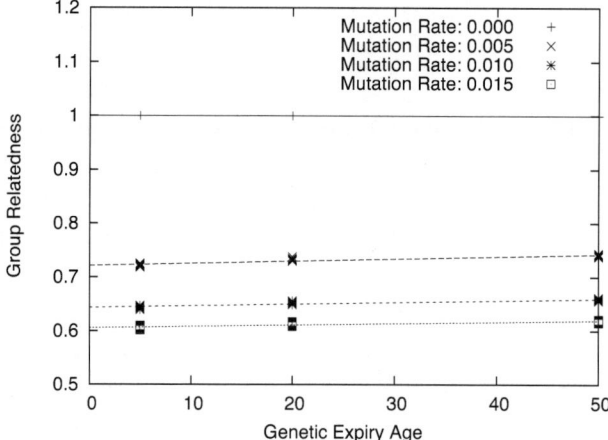

Figure 5.4: The group relatedness resulting from varied genetic expiry age and vulnerability mutation rate.

Experiment 2. Next, simulations were conducted to gauge the effects on the host populations of two disease split rates (0.10, 0.15) and different ge-

CHAPTER 5. EXPERIMENTS IN EVOLUTION

netic expiry ages (5, 20, 50) (see Figure 5.5). The global population density is a function of the two factors of group population sizes and the number of groups. These simulations were conducted with migration disabled, allowing the groups to become, and stay, extinct when disease epidemics wiped them out. In these simulation experiments, and the remaining aging experiments, the vulnerability mutation rate was fixed at 0.01, meaning that the group diversity was largely determined by genetic expiry age. We expected that groups with greater diversity (i.e., lower genetic expiry age) and lower disease split rates would be less susceptible to disease infection, and hence resist disease epidemics. In simulation runs conducted with greater disease split rates or genetic expiry ages groups became extinct, and the global population eventually died out. In simulation runs with lower disease split rates or genetic expiry ages groups resisted extinction and established an equilibrium with the disease population at varying levels of infection.

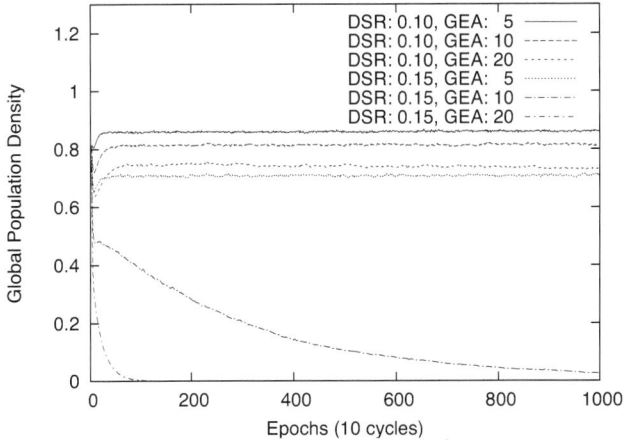

Figure 5.5: The global population density plotted over time, for varied disease split rates (DSR) and genetic expiry ages (GEA).

Experiment 3. For this experiment, migration was enabled, facilitating the expression of a group benefit for rapid aging, and the genetic expiry age gene was permitted to evolve. Simulations, 20 per parameter set, were run for 50,000 cycles, and the resultant average genetic expiry age at the end of each run was recorded. Figure 5.6 plots the final genetic expiry age against varied migration rate (0.001, 0.005, 0.01) and disease split rate (0.15, 0.13,

0.10). Figure 5.6 plots the corresponding proportion of all deaths attributed to the expiry age gene. The migration rate affects how often the groups attempt to reproduce themselves and the disease split rate affects the chances of a disease epidemic driving the group extinct. We expected that lower migration rates and greater disease split rates would select for faster aging. In simulations with greater migration rates and lower disease split rates, a greater genetic expiry age evolved. In some cases, the proportion of deaths attributed to genetic expiry, as can be seen from plot 5.6, became so low that there was only a slight selection pressure for longer life spans. If migration rate were reduced further, group selection would fail, as groups were unable to replace themselves before extinction. Likewise, if disease split rate were increased further, groups suffered disease epidemics causing extinction before being able to replace themselves. For all the following experiments the disease split rate was fixed at a probability of 0.15 and the migration rate at 0.001.

Calibration. We observed a great sensitivity to variation in the vulnerability mutation rate and disease split rate, whereas the results were relatively insensitive to variation in migration rate (Figure 5.6) and genetic expiry age (Figure 5.4). The insensitivity to the migration rate is a consequence of our requiring migrants to relocate to empty patches. As a result, when all neighboring group locations are occupied, the migration attempt fails, so pushing migration rates from medium to high often has no impact. We expect that a more subtle design, perhaps with migrants permitted to invade existing groups, would allow for more sensitivity to the rates of migration.

In Figure 5.4 group relatedness was plotted for varied fixed genetic expiry ages and vulnerability mutation rates. From it we see that the vulnerability mutation rate need only be altered slightly to get large changes in group relatedness. If the vulnerability mutation rate is decreased or increased too much, the aging rate will no longer evolve. This is because genetic similarity will be either too great, allowing diseases quickly to overrun host agents, or too small, with diseases unable to gain any foothold. A similar problem occurs with disease split rates. The differences in relatedness for varied genetic expiry age are fairly small (see Figure 5.4). Because of this, small changes in disease production greatly affect the simulation, possibly leading to either an unsustainable host or an unsustainable disease population. There is a delicate balance between the two different co-evolving populations. These sensitivity problems are most likely caused by the simplistic nature of the simulation. For example, in order to simplify

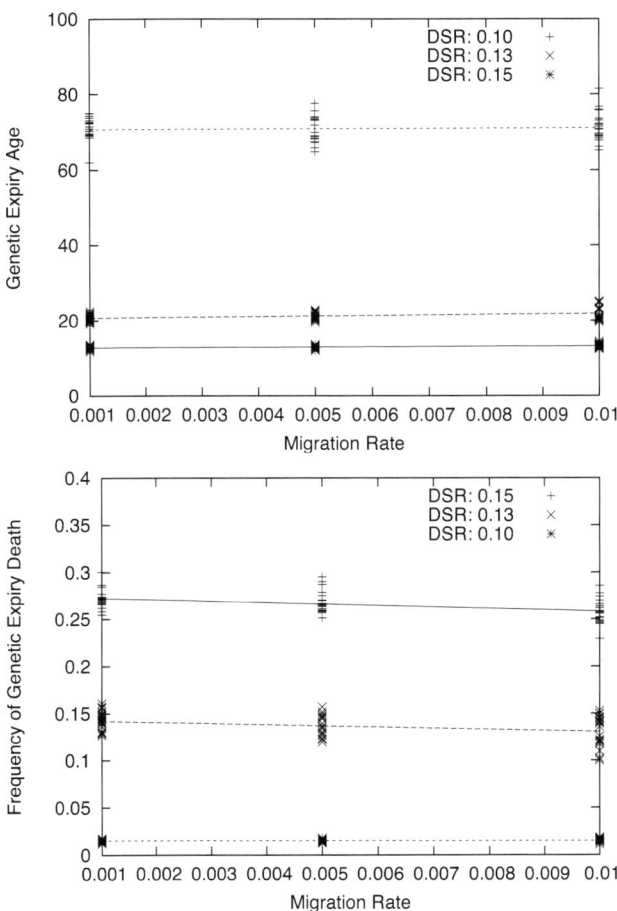

Figure 5.6: (a) The evolved genetic expiry age and the corresponding (b) proportion of all deaths attributed to genetic expiry plotted against migration rate and disease split rate (DSR).

the difficult simulation of co-evolving populations, disease virulence was not modeled and could not evolve, whereas in the real world the evolution of virulence can help diseases find an equilibrium with their hosts. Diseases which rapidly kill their hosts (and their energy source) either evolve restraint through group selection or else quickly drive themselves to extinction. In our simulations, they are simply reintroduced.

Figure 5.7 maps the benefit of aging and the relationship of some other simulation parameters. Rapid aging affects the group by increasing the rate of turnover which, coupled with vulnerability mutation rate, generates group mutation and thus increases group diversity (decreases group relatedness). Groups with increased diversity will better resist the invasion of diseases, which are generated according to the disease split rate, and so will be healthier and larger. Larger group sizes, together with higher migration rates, increase the rate at which new groups are pioneered, implying higher group fitness. Note that larger groups themselves feed back into more mutations and greater group diversity.

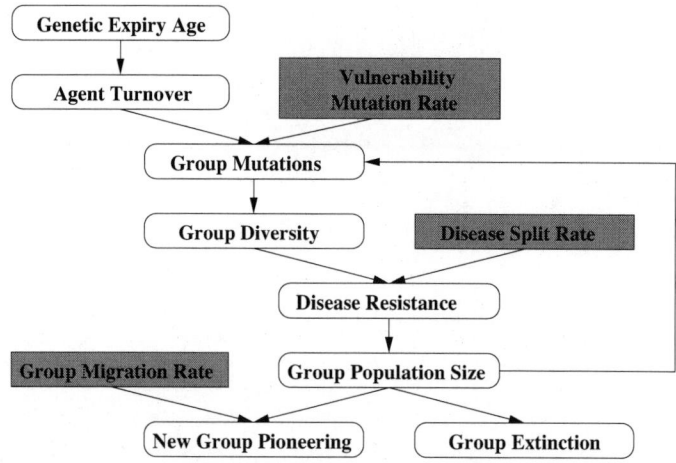

Figure 5.7: Mapping of the causal pathway between genetic expiry age and group benefits of new group pioneering and delayed extinction (simulation calibration parameters shaded). Edges only indicate effect and not the type of effect. Genetic expiry age is fixed in calibration experiments to determine effects on variables down the causal pathway.

CHAPTER 5. EXPERIMENTS IN EVOLUTION

Experiment 4. There are three different aging hypotheses we tested in this experiment:

- Our diversity hypothesis (DIV)
- The antagonistic pleiotropy hypothesis (AP)
- The mutation accumulation hypothesis (MA)

Each hypothesis was coupled with a simulation component which could be switched on or off for any given simulation. The DIV hypothesis required the existence of a disease population, which will exploit genetic relatedness. The AP and MA hypotheses required different genes to be enabled, namely, pleiotropic genes or the mutation accumulation death string respectively. We expected that if any given component were disabled then the associated hypothesis would fail to apply. In order to do a complete comparative study of these hypotheses, various simulations were conducted with all the different combinations of these components (see Table 5.1).

Table 5.1: Simulation Experiment Settings.

Experiment	Diseases	Pleiotropic Gene	Mutation Accumulator
NULL	✗	✗	✗
Diversity (DIV)	✓	✗	✗
Antagonistic Pleiotropy (AP)	✗	✓	✗
Mutation Accumulation (MA)	✗	✗	✓
DIV & AP	✓	✓	✗
DIV & MA	✓	✗	✓
AP & MA	✗	✓	✓
DIV & AP & MA	✓	✓	✓

Figure 5.8 (a) plots the evolution of the expiry age gene for each of the experiments in Table 5.1. Figure 5.8 (b) plots the corresponding proportion of deaths attributable to the gene. As can be seen in Figure 5.8 (a) our expectation that rapid aging would only have adaptive value with the disease population enabled (DIV) was fulfilled. In the other simulations Figure 5.8

(b) shows that deaths due to age drop to negligible levels, below 1%. In the simulations without diseases longer life spans were clearly selected for, with Figure 5.8 (a) showing no limit to the increase; individual selection reigned supreme, as there was no adaptive benefit to aging. The presence of the non-adaptive mechanisms (AP & MA), which also contribute to population turnover, cause the expiry gene to evolve to support longer life spans. Hence, in Figure 5.8 (a), the DIV & AP and DIV & MA results are greater than DIV by itself, and the joint DIV & AP & MA is greater still.

Figure 5.9 (a) plots the evolution of the AP gene. This is the value which determines the ratio of food energy directed to fertility and health (maintenance), where higher values equate to greater emphasis on maintenance and less on fertility. Figure 5.9 (b) plots the evolution of the MA gene. In order to simplify the representation of the MA gene string, this was measured as the first age at which a death mutation appeared. Both the AP and MA hypotheses rely on a decreasing probability of reproduction with age (see §5.2), thus we would expect a synergistic effect when the multiple aging mechanisms are enabled (i.e., DIV or AP or MA). From the figures, this appears to be the case. For example, the AP and the MA genes, in Figures 5.9 (a) and (b) respectively, were selected towards greater aging, i.e., lower AP and MA values, when both the AP and MA components were enabled. When the disease component (DIV) was enabled, we expected to see additional effects of the non-adaptive genes. First, since disease parasitizes its requirements from host health, it is of greater importance for the host to maintain this health store. This we expected to be a factor in the AP trade-off. It can be seen in Figure 5.9 (a) that when the disease (DIV) component was enabled, the AP gene evolved to put a much greater weight on health maintenance. Second, per our diversity hypothesis, when the DIV component was enabled, shorter life spans had an adaptive benefit, and this lessened the side-effect costs of the non-adaptive genes. This effect can be seen in both Figures 5.9 (a) and (b), where these runs evolved lower AP and MA genes. It is also noteworthy that in Figure 5.9 (b) there was very little difference between the DIV & AP & MA run and the DIV & MA run, with the DIV component diminishing the synergistic effect of the AP component on the MA gene.

As we noted earlier, Williams outlined the major difficulties with an adaptive theory of aging, one of which was that a high rate of accidental death implies an infrequency of aging deaths in the natural populations. In order to address this criticism all of our simulations had a high rate of accidental death, namely a probability of 0.1 per cycle. A high accidental

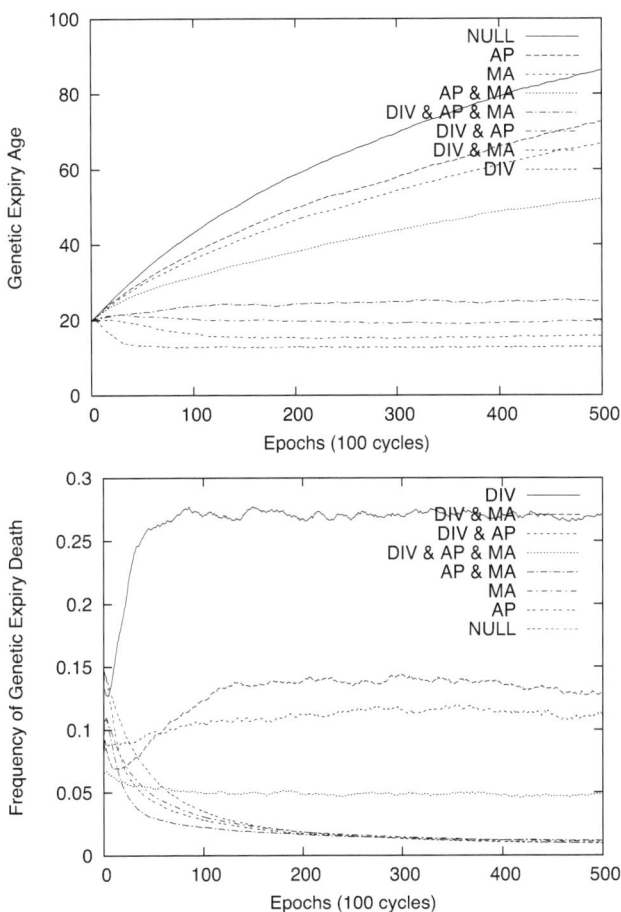

Figure 5.8: Plots over experiments in Table 5.1. Labels are sorted by final position. Figure (a) plots the evolution of the genetic expiry age gene which is present in all simulations. The evolution of aging occurs only in the DIV simulations, with diseases enabled. Figure (b) plots the proportion of deaths caused by the genetic expiry age gene of Figure (a).

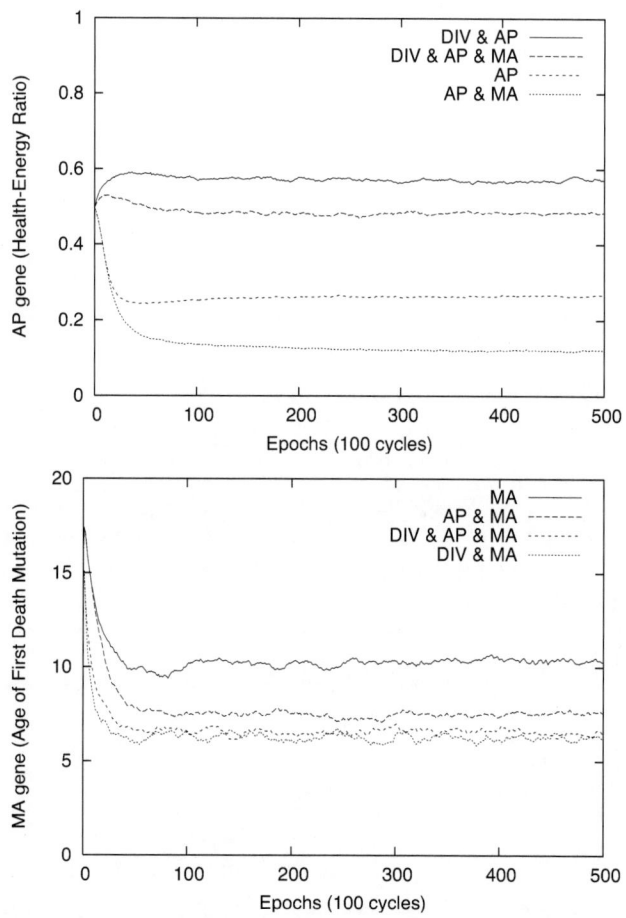

Figure 5.9: Figure (a) plots the antagonistic pleiotropy gene value, which determines the division of food to health and fertility stores — greater values signify less weight in health maintenance. Figure (b) plots the age index of the first appearance of a death mutation in the mutation accumulation gene string.

death rate was also required to test the antagonistic pleiotropy and mutation accumulation hypotheses. Even such high rates of accidental death, however, did not interfere with our diversity scenarios, which consistently and rapidly evolved shorter life spans regardless: there remained sufficient benefit in removing the lucky agents which survived to older ages. In fact, just as accidental deaths decrease the cost of aging in antagonistic pleiotropy and mutation accumulation experiments, it will also decrease the individual cost of aging in the adaptive experiments.

In summary, while antagonistic pleiotropy and mutation accumulation had clear impacts on the evolution of aging, in our simulations it was only the need for population diversity in the face of co-evolving diseases that could explain sustained short life spans in host populations. There may well be other adaptive explanations for the evolution of short life spans, as the work of Mitteldorf (2006) and Skulachev (1997) suggests, but the non-adaptive explanations of Williams and Medawar are unsuccessful.

5.2.2 Species Selection

Having demonstrated that our hypothesis of aging for the sake of diversity can be correct for a group-structured evolutionary environment, here we extend that work, demonstrating also its possibility in a scenario of species selection.

Simulation Design

We again used co-evolving populations of host and parasite agents, interacting within overlapping neighborhoods on a board, sharing food sources and potentially reproducing sexually. Table 5.2 provides an overview of the simulation parameters; selected ones are discussed in depth below. Unmentioned parameters have an interpretation substantially unchanged from the prior aging experiments.

Board: The simulation board consists of a 120x120 square grid of cells, in a torus. Each cell contains an occupant population, unlimited in size, and a food store. Each cycle the food store is replenished with new food, as determined by a normal distribution, and energy is recycled. The cell contents interact with the nine cells in the Moore neighborhood — recycled food is distributed, evenly, to neighboring cells, and agents feed, mate and migrate freely within their neighborhood.

Table 5.2: Simulation Parameters.

Parameter	Comment
Epoch Length	100 cycles
Run Length	100 epochs
Board Size	120 × 120 cells
Neighborhood Size	3 × 3 cells
New Food	$N(1,(\frac{1}{10})^2)$ units
Initial Health	20 units
Parental Health Investment	10 units/parent
Health Energy Overhead	1 unit/cycle
Max Health	80 units
Mature Health	60 units
Accident Rate	0.1
Parasite Generate Rate	0.0001
Signature Length	100 bits
Signature Mutator	0.005
Initial Expiry Gene	20
Expiry Mutator	$N(Expiry,(\frac{Expiry}{50})^2)$
Initial Airborne Gene	0.05
Airborne Mutator	$N(Airborne, 0.001)$
Airborne Co-ordinate Jump	$N(0,10)$

Host Agents: The host agents are the focus of the simulation. Each host agent occupies the cell into which they were born. Each cycle all host agents have an opportunity, in a randomly selected order, to eat and reproduce, after which they are tested for death conditions. The simulation maintains for each host agent its age, health, and chromosome. The health is incremented whenever the host agent successfully eats, which occurs each cycle, unless there is no food, or its health is either non-positive or greater than a maximum value. Health is also decremented each cycle for metabolism; it is also decremented by a parental investment amount whenever an agent reproduces.

Host agents are genderless, but reproduce predominantly sexually. After the agent has eaten, it is tested against a health threshold; if the agent has sufficient health, it attempts to reproduce. In addition to avoiding sui-

CHAPTER 5. EXPERIMENTS IN EVOLUTION

cidal mating, this forces agents to mature before reproducing, since new agents will lack sufficient health. When reproducing, the agent first checks its neighborhood for any mate requests by compatible agents (compatibility is discussed later). If one is found, the agents reproduce sexually and two offspring are created. If the agent fails to find a mate, but its health exceeds its maximum health threshold, it will reproduce asexually. The initial health of any child is the sum of parent health donations.

There are three causes of death. The host agent dies if:

1. its health falls to zero or below;

2. its age exceeds its genetically determined expiry age;

3. or it dies of accidental causes.

Having a closed ecosystem requires us to remove dead agents from the board and recycle any remaining energy held as health through the growth of new plant food.

Host Chromosome: The host agent chromosome contains:

- an expiry age gene;
- a mate compatibility bit-string signature;
- and a vulnerability (immunity) bit-string signature.

The role of the expiry age gene is the same as in our prior experiments. Mating compatibility, and species membership, is determined by testing whether the Hamming distance between the strings is greater than a fixed mating variance threshold (see Figure 5.10). The signature is inherited (via crossover) with a chance of mutation flipping each bit. This mechanism allows for the diversification of the mate signatures and thus the emergence of sexually isolated subpopulations, i.e., new species.

The vulnerability signature, as before, is the lock which parasites must open to infect hosts.

Parasite Agents: The parasites live off the host agent population. There may be an unlimited number of parasites living off a single host; however, if the host has non-positive health, it will die. Each cycle all parasite agents are transmitted to a randomly selected new host in their neighborhood, if

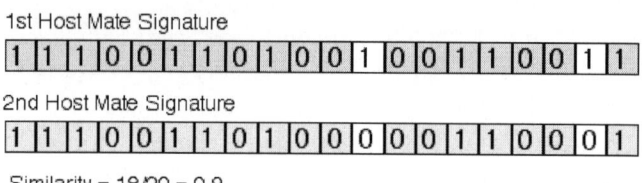

Figure 5.10: Example interaction between two host mate signatures. In this case, if the mate threshold parameter is set at a value lower than, or equal to, 0.9, the agents may mate, and thus are of the same species. (Note: all signatures in the simulation are 100 bits in length, not 20 as in this illustration.)

there is one. Occasionally a parasite will become airborne (with a probability determined by a gene in its chromosome) and is transmitted to a random cell on the board and a random host within that cell, if there is one. The co-ordinates of the destination cell are determined by sampling a normal distribution. If the new location is empty, the parasite dies. Becoming airborne provides the parasite population the opportunity to infect new populations.

When a transmission is successful, airborne or otherwise, the parasite agent attempts, twice, to parasitize health from its host and use it to clone offspring (one for each successful attempt), which will become active during the next cycle. Infection and reproduction are based on an interaction between the parasite and host chromosomes, discussed below. After the parasite has attempted reproduction, it dies. To ensure that the parasite population is never completely eradicated, there is a small probability a new parasite will be generated for every host agent updated.

Parasite Chromosome: The parasite chromosome has three components:

- an infection bit-string signature;

- a virulence bit-string signature;

- and an airborne transmission probability.

The probability of a successful infection is determined by the following function of the Hamming distance between the parasite's infection signature and the host's vulnerability signature:

$$P(infect) = \sqrt{\frac{Hamming\ Dist(infect, vuln)}{Signature\ Length}} \qquad (5.3)$$

Similarly, the probability of successful reproduction is determined by the Hamming distance between the parasite virulence signature and the host vulnerability signature in the same function (see Figure 5.11). Both signatures are inherited from the parent parasite with mutation.

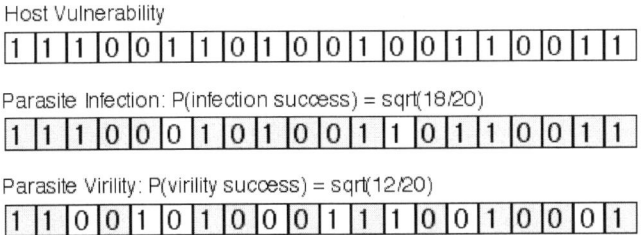

Figure 5.11: Example interaction between the host vulnerability signature and the parasite infection and virulence signatures. (Note: all signatures in the simulation are 100 bits in length, not 20 as in this example.)

Experiments

In order to explore the effects of species on the evolved aging rate, simulations were run with a variety of mate variance thresholds (see Figure 5.12) — a low threshold will allow host agents with greatly differing signatures to mate, reducing the number of species, whereas greater thresholds are more restrictive and thus produce a greater frequency of speciation. The resultant evolved genetic expiry ages and number of species for a range of mate variance thresholds are summarized in Figure 5.13.

From Figure 5.13 we can see that, as expected, the number of species present in the simulation increases as the mate variance threshold increases. We can also see that as the mate variance threshold increases, and thus the number of species and the strength of inter-species competition increase, the expiry age gene evolves for shorter life spans. It is noteworthy that even when there is only one species, and thus no inter-species competition, the expiry age gene evolves to an equilibrium; this must be based simply on a background kin selection pressure, since there is then no group or spe-

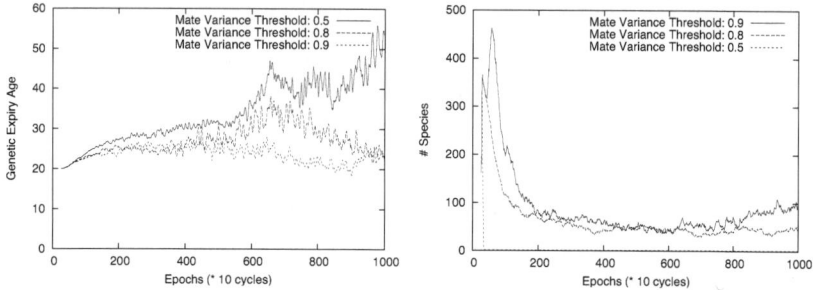

Figure 5.12: Figures tracking the evolution of (a) genetic expiry age and (b) number of species, for select cases of mate variance. Each of these plots represents a single simulation run. The resulting genetic expiry age and number of species are summarized in Figure 5.13.

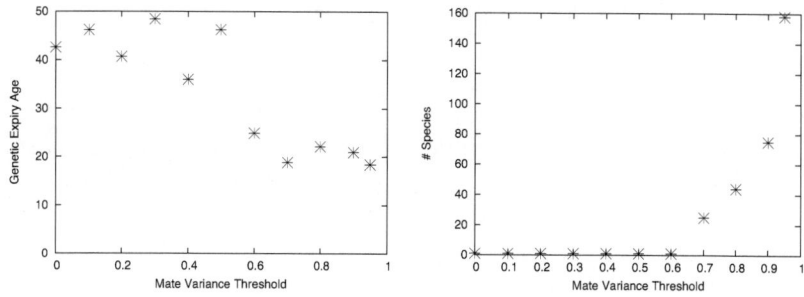

Figure 5.13: The (a) number of species and (b) evolved expiry age, after simulation runs were completed, with varied mate variance.

cies structure. The species selection pressure acting on top of kin selection drives the aging rates higher — i.e., it drives lifespans downwards.

We conducted additional experiments to analyse the effect of varying parasite virulence (see Figure 5.14). We would expect virulence to evolve via kin and group selection mechanisms in real parasite populations, however for simplicity we chose a parametric implementation of virulence, enabling these experiments. These results show that when the parasites are over-virulent, the host population is quickly driven to extinction and consequently the number of species drops to zero. When the parasites are under-virulent, they fail to maintain a foothold in the host population. In consequence of this, there are fewer dead-zones (where the parasites have killed off the host population), and the host population, which consumes the same

CHAPTER 5. EXPERIMENTS IN EVOLUTION

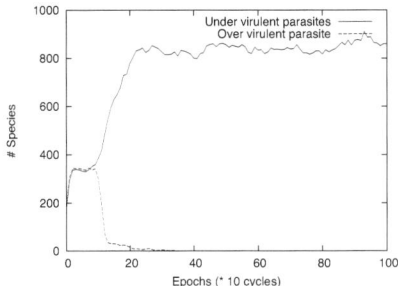

Figure 5.14: Experiments conducted varying the virulence of the parasites by varying the exponent in Equation 5.3 from 0.5 to 0.4 (under-virulent) and 0.6 (over-virulent).

amount of food regardless, spreads across the board, becoming less dense, which results in a greater rate of speciation.

5.2.3 Summary

As we mentioned in §5.2, Williams' main criticisms of the Weismann hypothesis, and more generally of adaptive theories of aging, were:

1. the fallacy of identifying senescence with mechanical wear;

2. the extreme rarity, in natural populations, of individuals that would be old enough to die of the postulated death mechanism;

3. the failure of gerontological research to uncover any death mechanism;

4. and the difficulty of conceptualizing how such a feature could be produced by natural selection.

We conclude by addressing each of these criticisms, except the third which is outside the scope of this paper. (However, we note that this criticism applies equally to the alternative theories.) The first of these criticisms concerns particularly Weismann's hypothesis of aging. We do not presuppose any relation between senescence and age: apart from the increased probability of death with age, the host agents in our simulation retain a constant ability to reproduce throughout their lives. Our diversity hypothesis, like Weismann's, holds that aging is adaptive with a group benefit of making room for the young. However, we formulate this as: populations which experience aging turn over faster and therefore have greater diversity. In order to explore the second criticism, we included a high rate of accidental death in our simulations. High rates of accidental death in no way impeded the evolution of faster aging rates. The fourth criticism was formulated before the papers of Hamilton, Price and Maynard Smith laid out selection mechanisms which support altruistic adaptations (see §5.1). Both kin selection and group selection, and even more the two operating together, are fully capable of explaining the evolution of altruism, including the evolution of faster rates of aging.

In §5.1 we outlined an argument that group selection is supervenient on kin selection. The Price equation requires that for the group selection of altruistic genes there be a positive association between group fitness and frequency of altruistic genes, which is likely to arise via kin selection. The reliance of a model on kin selection can be tested by the removal of kin associations, i.e., ensuring that there is no correlation between the locations of the parent and child. When we performed this test in many of our aging experiments, the removal of kin selection resulted in the collapse of group selection.

Our ALife simulations demonstrate that groups with shorter life spans do indeed have a greater diversity. This diversity has a fundamental effect in the co-evolutionary arms race between host agents and the disease agents which parasitize them. Host populations with shorter life spans, and therefore greater diversity, are less likely to experience a disease epidemic and can allocate more energy into sustaining a greater population of hosts. We implemented a group selection mechanism, where groups with larger host populations were more likely to produce successful migrant parties to found new groups and conducted a comparative study with two other popular non-adaptive explanations of aging. The evolution of aging genes occurred readily in the simulations demonstrating the diversity hypothesis, although, when available, genes with other positive side effects were selected over these simple genes.

In sum, our simulations show the viability of an adaptive account of accelerated aging as a consequence of a combination of group and kin selection, including species selection, incidentally demonstrating the utility of ALife simulation methods for developing and testing evolutionary hypotheses.

CHAPTER 5. EXPERIMENTS IN EVOLUTION   135

5.3 Suicide as an Evolutionarily Stable Strategy

We now turn to the somewhat surprising idea of evolving suicide.[4] Since a genetically controlled suicide has a fixed and strong tendency to eliminate itself from the gene pool, it is counterintuitive to think it could evolve. However, in special circumstances suicide can be altruistic, freeing up resources for one's kin, and in those circumstances it can evolve. We decided that this combination, the possibility of suicide being altruistic under special circumstances together with it being the obvious candidate for a behavior most unlikely to evolve, made suicide an excellent vehicle for demonstrating the power of kin selection to fix altruistic behavior in a population.

The Evolution of Suicide

Suicide may have an adaptive explanation, but there are several hurdles we must first overcome before developing one. The simplest evolutionary approaches focus on individual survival and reproduction, but individuals who commit suicide put an end to both; suicide is thus impossible to explain at this level. Instead, we need to take into account the effect on genetic relatives. Still, suicide poses a problem even for kin selection, since individuals who commit suicide also prevent themselves from helping kin. For suicide to be adaptive, the individual's continued existence must somehow be a burden overall to its kin.

de Catanzaro (1986) has developed a simple mathematical model using kin selection and the idea of burden to clarify when suicide may be adaptive. The model, which has received some modest empirical support (for example, see Joiner Jr et al., 2002, Brown et al., 1999, de Catanzaro, 1995), can be summarized by the following equation:

$$\psi_i = \rho_i + \sum_k b_k \rho_k r_k \qquad (5.4)$$

where i is an individual and k its relatives, ρ_i (or ρ_k) is the remaining reproductive potential of individual i (or k), b_k is the coefficient of benefit (or cost) to the reproduction of k given the continued existence of i, and r_k is the coefficient of genetic relatedness between i and k. ψ_i, then, describes the adaptive value of i's continued existence. Assuming an ideal adaptation to the environment, $\psi_i < 0$ dictates that the individual will end

---

[4]Note that in this section we describe experiments more recent than our original work, (Mascaro et al., 2001). This work varies in detail from the original work and was done in NetLogo. The simulation is available from the book web site.

its existence, $\psi_i > 0$ dictates that it will continue its existence, and $\psi_i = 0$ indicates adaptive indifference. The quantity $\sum_k b_k \rho_k r_k$ describes the total benefit or burden the individual has on its kin. When the quantity is sufficiently negative, then $\psi_i < 0$ and the individual will commit suicide, if it has such a choice and its behavior is optimally adaptive.

It might be argued that suicide can evolve even when an individual is not a burden on its kin, so long as suicide boosts the reproductive potential of kin sufficiently. But this case is already covered by the equation — we can observe that *not* boosting the reproductive potential of kin by remaining alive is a kind of burden on the reproductive potential of kin (by reducing reproductive opportunity), and so represented by b_k in Equation 5.4.

While suicide may be *one* solution to the evolutionary problem of burdening one's kin, there may be other solutions. Take the example given by de Catanzaro (1986) of an infirm 75 year old female who is supported by her financially pressed daughter with children of her own. Clearly, the burden put on the daughter is not necessary — there are any number of unburdening interventions possible, including social welfare. Thus, suicide is just one of several strategies that may evolve to deal with the burden of this kind of case. This is also true of many other situations (perhaps all) in which suicide can be a solution to the evolutionary problem of burden on kin; in such cases, suicide is a sufficient adaptation, but hardly a necessary adaptation.

If there are multiple strategies that can evolve, it is not clear whether suicide specifically will do so. However, if suicide can evolve *quicker* or more easily than other strategies, then it will neutralize the evolutionary advantage of those other strategies. That is, once suicide is common, natural selection will no longer act to promote the spread of other strategies for dealing with such burdens.

Thus, if suicide is to become evolutionarily stable, then either 1) there must be no other strategies for dealing with burden, or 2) suicide must evolve quicker or more easily than the alternative strategies for dealing with burden. In our simulations, the burden that each agent places on other agents involves the consumption of common food and participation in fruitless mating. When the burden is severe, agents may evolve to refrain from eating and mating, or, alternatively, suicide may evolve — spreading more quickly, since it is a readily available act that puts a certain end to these other acts. We now describe a simulation that puts these ideas into effect.

We should note first, however, that it turns out that Brinkers and den Dulk (1999) developed an evolutionary suicide simulation shortly before

CHAPTER 5. EXPERIMENTS IN EVOLUTION

we did and also using a group selection model. In their simulation, agents live on 10x10 islands, separated by canals one cell thick. Agents reproduce asexually, with children born alongside their parents, and eat the food available on their island. Normally, migration is implemented in such group-selection models by having some probability of migration between groups. Here, however, agents can only cross between islands if an agent sacrifices itself to form a bridge with its own body. Their agents did then evolve to sacrifice themselves, albeit at a very low rate. When the authors turned off kin selection, using a method equivalent to our adoption queue of §5.1, the rate of self-sacrifice dropped substantially but did not disappear completely. However, suicide was then supported in their simulation only by mutation; as the mutation rate was further reduced, self-sacrifice declined until the population died out. In our simulations suicide evolves in a much more robust fashion and, arguably, under more biologically plausible and relevant circumstances.

5.3.1 Simulation Design

A good place to begin our description of the suicide simulation is with the table of parameters given in Table 5.3, which also serve as a convenient reference for the setup of the simulation.

Food Distribution Function

The *fdf* can either be a *seasonal* or a *constant-rate* food supply function. A seasonal food supply means we generate food for the agents according to a sine wave over time. The sinusoid has a period of 120 cycles, and a range of 0 to 40 food units (and thus it has a magnitude of 20 food units). The seasons of interest are the half cycles corresponding to periods of drought and plenty. In contrast, a constant-rate food supply involves generating food at a constant amount per cycle. These two *fdf*'s allow us to contrast runs in which agents live in static environments with those in which agents are regularly put under pressure, such as in droughts.

Group Selection

The simulation employs a group selection model similar to that described for our aging experiments — groups occupy cells, are driven extinct over time, and occasionally have the opportunity to pioneer new cells. Group extinction is, predominately, driven by drought conditions. Group pioneering

PARAMETER	VALUE(S)
Food distribution function	Seasonal (sinusoid with period of 120 cycles and magnitude of 20 food units) or Constant rate (20 food units per cycle)
Actions	Eat, Reproduce, Migrate, Suicide
Observations	Self-age, Self-health, drought-condition
Board size	15x15
Maximum entities per cell	Unlimited
Eating neighborhood	1x1
Observation neighborhood	Local
Migration rate	0.0005
Migration neighborhood	3x3
Initial number of agents	225
Initial health	30
Agent age limit	50
Health from food	5 health units
Health required for mating	30 health units
Parental investment	10 health units
Action probability mutation	Initial mutation rate: 0.001, Meta-mutation rate: $N(0, 0.0001)$
Condition value/operator mutation	Initial mutation rate: 0.001, Meta-mutation rate: $N(0, 0.0001)$

Table 5.3: Parameters of the altruistic suicide simulation.

CHAPTER 5. EXPERIMENTS IN EVOLUTION

is driven by a small probability that individuals will migrate to a neighboring cell. If groups with higher suicide rates tend to have greater longevity or fecundity, this will create a positive group selection pressure that will counteract the individual selection pressure against suicide.

Agent Updating

Each cycle all agents are updated in a random order. During the update, each agent will have the opportunity to eat, reproduce, migrate and die (either by suicide or natural causes). The agent may perform all these actions in a cycle if it meets their preconditions. Of these actions only suicide is determined by an agent decision function.

Eating. The agent eats from its cell. Agents eat every cycle, if food is available. The amount of energy transferred is the food available capped at a maximum.

Reproducing. Agents reproduce asexually. As soon as the agent meets the reproductive health requirement, an offspring is created and placed on the board in the same location. The initial health of the offspring is deducted from the parent.

Migrating. Migration occurs with a fixed probability. It causes the agent to be relocated to a random cell in the 9 cell neighborhood. As this is the only movement that occurs in the simulation it provides for a group selection effect — via migration, groups may spawn new groups in empty cells.

Suicide. The implementation of suicide is simple: an agent that chooses suicide will immediately die.

Death. There are three causes of death in the simulation. Agents die if their age exceeds an expiry age, or if their health falls to or below zero, or if they commit suicide. In all cases, the agent is removed from the board and any remaining health is recycled back into the cell's food store.

The Agent Decision Function

The decision function for the simulation, that is, that organ of the agent which maps observations to acts, is a decision tree, as described in §4.6.2.

In this simulation the decision function has only two actions, suicide or don't suicide. The isolation of the suicide gene was chosen to ensure that suicide will not "piggy-back" on another action (see §5.3.5).

Observations

Agents can observe and condition their actions upon:

- Self-age: their own age scaled according to the maximum age
- Self-health: their scaled health
- Drought condition: the value of *fdf* at the time step divided by the number of agents on the cell, i.e., the number of food units per agent

5.3.2 Basic Demographics and Orientation

Figure 5.15 covers some of the relevant demographics for a typical suicide simulation. Population sizes are larger and less variable in the constant-food scenario, as one would expect. In the seasonal food simulations, average age is positively correlated with the amount of food generated. However, just before the peak food supply, and for a little while after the peak, the average age drops dramatically. This corresponds to a large number of births at this point in the cycle (the number of births peak just after the peak in generated food). In general, the average health is higher in the constant-rate food simulations, as we would expect, since life should be easier there. In the seasonal simulations, average health is higher after a peak in food than prior to it. Group age is the number of cycles since the cell was last empty, i.e., the time since group's founding. In the constant-rate food run the median group age steadily increases, since groups don't go extinct. In contrast, in the seasonal food run the median group age plateaus.

5.3.3 Experiment: The Evolutionary Stability of Suicide

For a strategy to be an ESS, it must be capable of resisting invasion from any mutant strategy. For some strategies we can use analysis to determine whether they are evolutionarily stable. This is not an option here. As with most ABM simulations, ours would be far too complicated to analyse in the required way. Instead, we have settled the issue experimentally. That is, we found that in our simulations a mixed strategy of suicide resisted invasion from mutant strategies avoiding suicide.

CHAPTER 5. EXPERIMENTS IN EVOLUTION

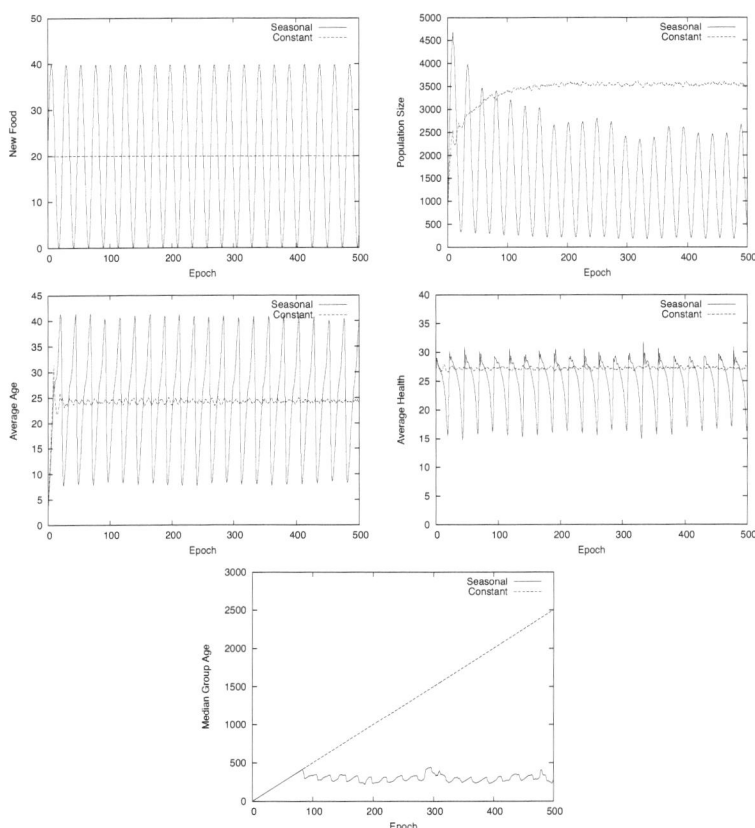

Figure 5.15: From a pair of typical seasonal and constant-rate simulation runs: (top left) food generated over time for the constant-rate and seasonal food supply simulations; (top right) population numbers over time; (middle left) average age over time; (middle right) average health over time; (bottom) median group age.

One way in which we can establish that suicide is an essential part of an ESS is by comparing two suitably similar situations in which evolutionary change has settled, that is, two situations in which mutation is expected to reintroduce suicide at the same rate. Thus, if we have two situations with suicide rates s_i and s_j, and a rate of reintroduction due to mutation s_m for both situations, then we must have:

$$s_i \geq s_m \tag{5.5}$$

$$s_j \geq s_m \tag{5.6}$$

If we can then show that the rate of suicide is greater in one situation, say, $s_i > s_j$, then $s_i > s_j \geq s_m$ or:

$$s_i > s_m \tag{5.7}$$

Recall that the definition of an ESS is that of a strategy that is capable of resisting invasion from any competing strategy. Here, s_i resists invasion from strategies containing suicide rates lower than $s_i > s_m$ — hence, s_i is part of an ESS.

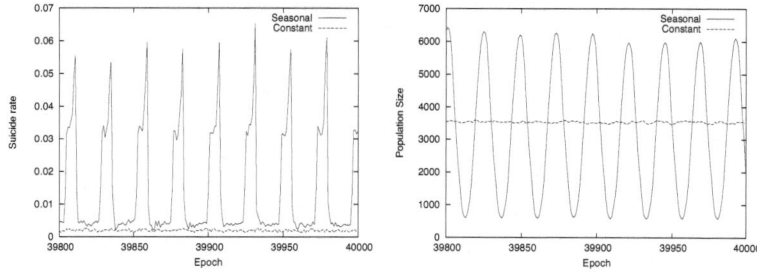

Figure 5.16: Suicides as a proportion of all acts in constant and seasonal food supply (left); population (right).

As we noted in §4.7, our simulations can collect an exact statistic for the suicide rate: the number of suicides per turn per agent, yielding the rate of suicides chosen as a proportion of all actions chosen. Suicide numbers per epoch were used for our statistical tests. Figure 5.16 shows a typical graph of the suicide rates for paired simulations of seasonal and constant-rate food supply (i.e., simulations with identical parameters other than food supply). With seasonal food, the suicide rate stabilizes into a cyclical pattern in short order (the early stabilization is not shown in Figure 5.16). The peaks in the suicide rate correspond to the troughs in population size.

The graphs clearly show that the suicide rate is above the reintroduction rate due to mutation. The suicide rate in the constant food run set, s_c,

CHAPTER 5. EXPERIMENTS IN EVOLUTION 143

must be greater than or equal to the reintroduction rate due to mutation, s_{m_c}, since mutation was active at the same level in both simulations. Further, the average peak suicide rate in the seasonal food run set, s_p, clearly satisfies $s_p > s_c$. In any case, over a set of twenty paired simulations a one-tailed significance test showed that the average seasonal-food suicide rate was statistically significantly greater than the average constant-food suicide rate. While mutation was constantly presenting our simulation with strategies having a smaller rate of suicide, in the seasonal simulations these did not replace strategies which showed a greater tendency to commit suicide during times of drought.

5.3.4 Possible Causes of the Evolutionary Stability of Suicide

While an adaptive explanation of the stability of suicide seems inevitable, we will now look at what the causes of suicide in the seasonal simulations may be, by examining the conditions under which suicide forms part of an ESS. Identifying such conditions is a part of any experimental analysis of the strategy in which suicide participates. For example, were we to discover that suicide occurs more frequently in droughts, then suicide is clearly part of an ESS which specifies that an agent will suicide with higher probability in droughts. By identifying such conditions, we can gain an understanding of the kinds of situation in which suicide is adaptive.

Properties of Agents that Commit Suicide

Figure 5.17 compares the distributions of the observed food availability, ages and health for those that commit suicide with the same distributions for the entire population in a the seasonal food run set. In looking at the two distributions for food density per agent, we can see that suicides are far more likely to be triggered by drought conditions, i.e., when food is scarce. Furthermore, we can also see that older agents are much more likely to commit suicide. Older agents tend to be more healthy (which explains the difference in the health distributions), since health does not deteriorate with age in our simulations; but their chances of future reproduction will be low, since they will soon encounter the reaper in the form of their maximum age (see Table 5.3).

144   CHAPTER 5. EXPERIMENTS IN EVOLUTION

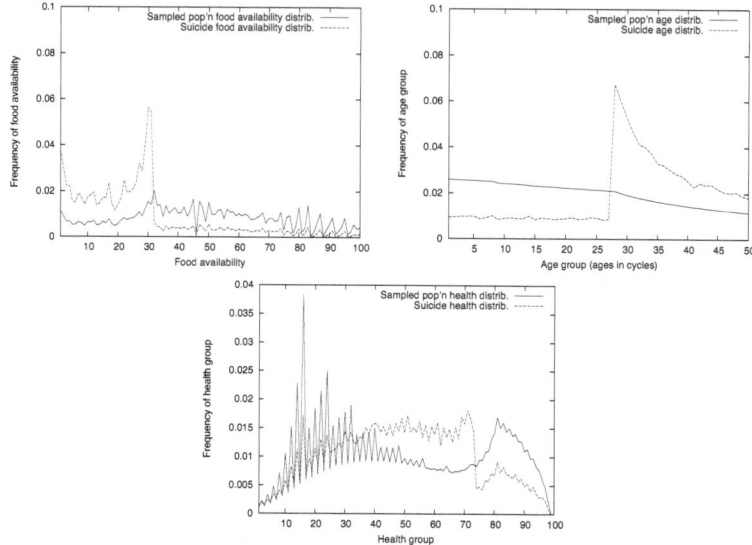

Figure 5.17: Distributions of food density per agent (top left), ages (top right) and health among agents that committed suicide and among the general agent population (bottom).

5.3.5 Alternative Explanation: Mutation Accumulation

As we saw above, Medawar argued that a mutant gene which affected survival early in life would have a higher net effect on fitness than one which affected survival late in life, when the reproductive value was relatively low. Thus, harmful mutations, which we may experience as aging, affecting the organism late in life, would not be as effectively weeded out by natural selection and would tend to accumulate.

This explanation of aging can be extended to a non-adaptive explanation of suicide. Suicide differs from aging in that it is a behavior conditioned on self and environmental observations. Certain observations, as of high age or of extreme drought conditions, are less likely and therefore the behaviors triggered by them will be subjected to a lower selection pressure. In our simulation, it can be seen (Figure 5.17) that older agents are much more likely to commit suicide, which directly corresponds to Medawar's hypothesis. However, the evolution of the observational conditions for the basis of choosing suicide, i.e., a distinct preference for drought conditions and for high age, shows that there remains some adaptive value. In the constant-rate food runs, where mutation accumulation is every bit as active as in the seasonal food runs and where we expect suicide to be valueless, the behavior decision tree governing suicide evolves to a single leaf node, making no distinctions whatsoever.

5.3.6 Summary

Thus we find that suicide need not be pathological (as the early de Catanzaro held; see, for example, de Catanzaro, 1981), nor outside the scope of evolution. The evolutionary mixed strategy employing suicide in our simulations provided a group fitness benefit. Suicides were conditioned upon at least low food density and high age, implying that removal of the agent from the simulation allowed other group members, who were more likely to be able to reproduce in the future, better access to food, and so suicide produced greater average health within the group, allowing remaining agents to better cope with the drought. In these, admittedly special, circumstances suicide has adaptive value. Those circumstances were, of course, *designed* to provide a fitness benefit for suicide, with reliably repeated situations of environmental stress. Whether, and how far, the benefit of suicide extends beyond the torus world of these simulations is an open question. Clearly, however, evolved altruism, of which suicide is only an extreme example, is widespread, both naturally and virtually.

5.4 The Evolution of Parental Investment

Evolutionary biologists have debated **sexual selection** since Darwin, who used it to explain the female choosiness and male competitiveness he observed in most species. In the middle of the last century Bateman (1948) gave Darwin's observations firmer support with his experiments with Drosophila (fruit flies). Bateman found that male Drosophila mate with as many females as possible, while female Drosophila are more restrained, mating with just one or two males. He speculated that this difference in behavior is due to the difference in gamete sizes of the two sexes, with the female's being much larger and therefore more costly. Further, he suggested that the different behaviors would lead to differences in the variability of reproductive success — specifically, that females would have roughly equal success, but some males would do very well at the expense of others who would do poorly.

Two decades later, Trivers took inspiration from Bateman's work and generalized the idea of dimorphic gamete cost to the idea of dimorphic *parental investment* (Trivers, 1972). According to Trivers, parental investment is "any investment by the parent in an individual offspring that increases the offspring's chance of surviving (and hence reproductive success) at the cost of the parent's ability to invest in other offspring" (Trivers, 1972, p. 139). Parental investment covers any cost involved in looking after an offspring, whether in gamete production, gestation or care after birth. The definition specifically omits any effort put into attracting mates or competing with members of the same sex. That kind of effort is called mating investment; we are often interested in the interaction between this and parental investment (for example, see Heath and Hadley, 1998). Biologists have also identified other related concepts, such as parental care (which occurs strictly after birth); see Clutton-Brock (1991) for a detailed review.

The concept of parental investment enters into the explanation of many cases of sexual selection and the evolution of reproductive strategies. Trivers has used it to explain Darwin's observations of female choosiness and male competitiveness in species where females are the higher investors (Trivers, 1972). He has also used it to explain a parent's ability to vary offspring sex ratios in some species (Trivers and Willard, 1973), and the periods of conflict that arise between a parent and its offspring during weaning (Trivers, 1974). Others have applied it to infanticide and abortion (Hrdy, 1979, Lycett and Dunbar, 1999), the greater rate of child homicide amongst stepfathers and boyfriends (Daly and Wilson, 1994), and the perpetration of rape being principally by males (Thornhill and Thornhill, 1983, Thornhill and Palmer, 2000).

It is no coincidence that all of these uses of the parental investment concept are of ethical interest. Parental investment is a *social* concept, in which one individual makes sacrifices for the benefit of others. An obvious, but odd-sounding, utilitarian point is that parental investment is normally an ethical act, contributing to social welfare.

We would not normally consider parenting behavior in an ethical context because it seems essential to who we are and what we do — in other words, it is an integral part of our evolved psychology. Nevertheless, we do start considering the ethical implications of parenting behavior when dealing with less usual cases. We will consider two of these cases — namely, rape and abortion — in the next chapter. In the case of rape, we consider how the minimal investment required of males after rape (in contrast to females, who gestate) can lead to sexually dimorphic rape rates, as well as dimorphism in other behavior. In the case of abortion, we examine how unpredictable environments and different quantities of parental investment affect females' decisions to terminate such investments. Both of these investigations involve exploring the consequences of sexually dimorphic investments.

These issues raise the prior general question of how sexual differences in parental investments arise in the first case and the further specific question of why parental investment even after birth is usually greater for females. It is answers to these questions that we are now concerned. Biologists have suggested several possibilities for the origin of dimorphic parental investment. Trivers (1972) suggested that prior differences in investment can cause further differences in investments to evolve. Dawkins and Carlisle (1976), while identifying a fallacy in Trivers' reasoning — the idea that bad initial investments make further bad investments good, suggested an improved hypothesis: that the sex that *can* quit investing first *will*. Trivers (1972) also suggested that males who were less certain of their parentage would invest less. Finally, Williams (1975) noted that the sex that remained with the offspring due to some **pre-adaptation** would be in a position to evolve parental care.

Interestingly, each of these hypotheses depend on some prior difference between the sexes. Trivers even suggested the answer might lie in prior sexual differences being passed from species to species — leaving the ultimate explanation in a chance event in early evolution (Trivers, 1972). Unfortunately, this leaves the occasional reversal of parental care roles between male and female entirely unexplained. And, in any case, the continued maintenance of substantial **sexual dimorphism** in investment is certainly itself not simply an accidental process, but subject to positive selection pressure, which we would like to understand.

In this chapter, we will explore each of the above hypotheses. We will apply ALife simulations in scenarios matching the conditions of each hypothesis and check how well the hypotheses' predictions concur with the results. In some of the simulations, agents will directly evolve a parental investment; in others, agents evolve a period of parental care. As we will see, there is potential explanatory value to all of the hypotheses, barring (unsurprisingly) Trivers' original fallacious hypothesis. In the next chapter, we will take dimorphic investments as a given and explore their consequences for rape and abortion.

Variable	Definition
tpi	Total parental investment
mpi	Minimum parental investment
$pcpi$	Per-cycle parental investment (after birth)
epi	Evolvable parental investment
eit	Evolvable investment term

Table 5.4: Variables representing parental investments in the experiments.

5.4.1 Simulation design

Variable Parameters

Parental investment. Parental investment, pi, and the parental investment term, it (i.e., time period of investment), are the most important agent properties here. Several simulation variables represent investment, and these are summarized in Table 5.4. Note that subscripts always denote sex; for example, tpi_f represents the total parental investment made on average by each female.

Agents can either evolve their parental investment or their investment term, but not both. Genes determine what (or for how long) an agent invests in offspring, coded in each agent for both male and female investments. A child inherits both genes from a randomly chosen parent, but only expresses the gene corresponding to its own sex (of course). These genes are mutated at a rate determined by a mutation gene that is also a part of the chromosome. This allows the system to meta-mutate these mutation rates, allowing for adaptive mutation levels to evolve.

Fixed Parameters

The simulations were run for 7000, 20,000 or 40,000 cycles in groups of 30, 50 or (for paternal uncertainty) 101 runs. In most of the simulations, 50 pieces of food were generated each cycle. Each piece of food provides roughly 70 units of health to any agent that eating it. In all these experiments, agents required a minimum of 200 health units before they could reproduce.

Distinguishing Females and Males

In biology, the female sex is defined as that which has the larger gamete. In these experiments, we kept every non-experimental aspect of the sexes

CHAPTER 5. EXPERIMENTS IN EVOLUTION   149

equal, including any gametic investment. Therefore, instead we define the
female sex as that which evolves to invest more or that which started the
simulation with a greater initial or minimum investment.

5.4.2 Prior Investment Hypothesis

Trivers (1972), in his seminal essay on parental investment, implied what
at first may seem a plausible hypothesis. Namely, that the sex that commits
the most prior investment has the most to lose — and thus is the sex more
likely to evolve further investment. If correct, this implies that an arbitrary
initial difference in parental investments may lead to greater differences of
the same kind. We call this the *prior investment hypothesis*.

This hypothesis was criticized by Dawkins and Carlisle (1976), who
pointed out that it involved fallacious reasoning — of the sort used to justify continued spending on a project based on how much has been invested,
rather than what future investment will likely return, in violation of the
Bayesian principle of maximizing (future) expected value. They illustrated
their point with the then topical example of government spending on a supersonic airliner based on past spending, and the fallacy is now often referred to as the 'Concorde fallacy'.

It is important to distinguish this hypothesis from the claim that the
past investment that an individual makes is an indicator of their future investment. While the prior investment hypothesis is a causal conjecture, the
latter claim is a probabilistic conjecture.

Experiment: Dimorphic Investments as Initial Conditions

To test the prior investment hypothesis, we ran simulations under two configurations — the first described here, the second described shortly. In the
first configuration, an agent can invest in just one way, by transferring some
of its health to its offspring at birth, which we call the *total parental investment* (or *tpi*). As noted earlier, there is a *tpi* for each sex — tpi_f and tpi_m
— the genes for which each agent inherits from a randomly chosen parent.
The test of the hypothesis is then quite simple: we initially set $tpi_f > tpi_m$
for all agents at $t = 0$ (i.e., $tpi_{f,0} > tpi_{m,0}$), and then allow them to evolve.
If the prior investment hypothesis holds, then $tpi_f - tpi_m$ measured late in
the simulation should be greater than the same difference at the beginning.

Results. Some of the results from this test are shown in the graphs in the
left column of Figure 5.18. For both graphs, $tpi_{m,0} = 0$, while in the first

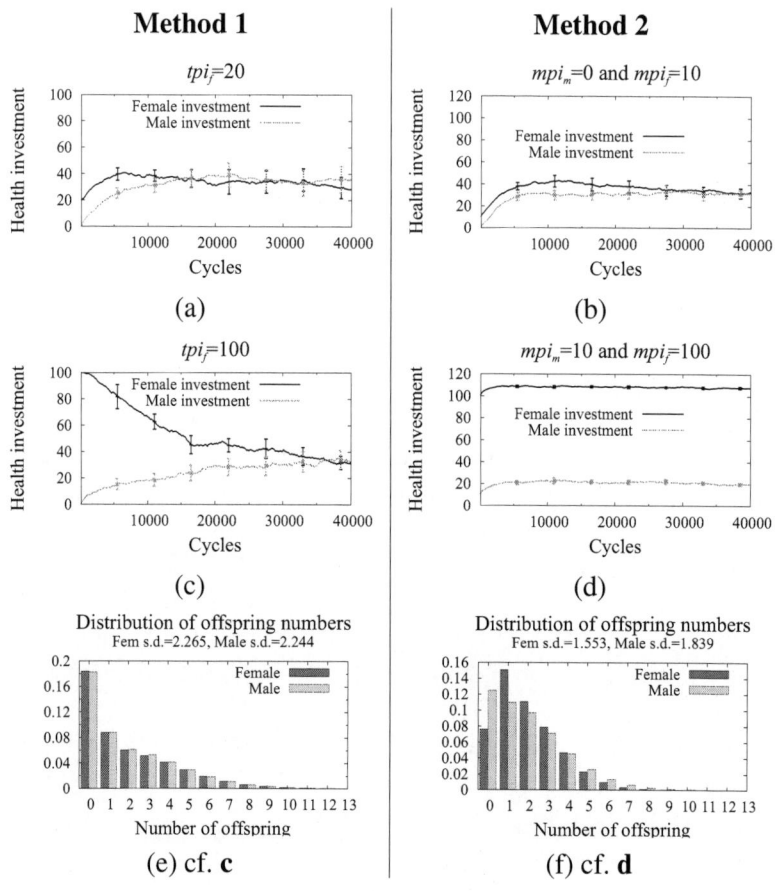

Figure 5.18: [Left column, Method 1] (a) Evolved *tpi* made by males and females when $tpi_{m,0} = 0$ and $tpi_{f,0} = 20$ and (c) $tpi_{m,0} = 0$ and $tpi_{f,0} = 100$. [Right column, Method 2] (b) Evolved *tpi* for simulations that set minimum parental investments for each sex of $mpi_m = 0$ and $mpi_f = 10$ and (d) $mpi_m = 10$ and $mpi_f = 100$. [Both] Distributions of reproductive success by sex for the simulations in (e) **c** and (f) **d**.

graph, $tpi_{f,0} = 20$ and in the second graph, $tpi_{f,0} = 100$. The graphs show the *tpi* that evolves for both males and females over time. It is quite clear from these graphs that, regardless of the initial settings for *tpi*, $tpi_f - tpi_m$ does not evolve to be greater than it was at first. Indeed, quite the opposite happens — that is, sexually dimorphic investment disappears entirely. We also produced other run sets (not shown in the figure) with different initial settings for tpi_f and tpi_m, ranging from 0 to 100 with the same result.

CHAPTER 5. EXPERIMENTS IN EVOLUTION

	Eat	Mate
$tpi_{f,0} = 100$	-0.54% (81.7%)	0.54% (15.6%)
$mpi_f = 100$	7.4% (76.1%)	-6.9% (20.5%)

Table 5.5: Female minus male action rates; female action rates in parentheses.

As we mentioned above, Bateman (1948) contributed a key idea to parental investment theory: that the sex that invests more will evolve to have less variance in its reproductive success. In contrast, we would expect there to be no difference in reproductive variability if both sexes invest equally. We checked this prediction in Figure 5.18e, which was taken from the last ~7000 cycles. The graph is a frequency distribution of the number of offspring agents have, divided by sex. As we can see, the distributions are nearly identical. While the distributions are significantly different on a chi-square test, ($\chi^2 = 161$), that is achieved only with a very large sample size (> 100000) and the Kullback-Leibler (KL) divergence between the distributions is negligible (7.5×10^{-5}).

We can discover whether sexually dimorphic behavior is evolving by looking at Table 5.5. The top row shows the female minus male difference in action rates that evolves for the $tpi_{f,0} = 100$ run set (the numbers in parentheses are the female action rates alone). As we can see, females evolve to eat ~82% of the time, while they evolve to mate ~ 16% of the time (the remaining ~2% is due to resting). In fact, there is an initial rapid move toward dimorphism in both eating and mating (not shown), with males mating more, and females instead eating more. This is almost certainly due to the vast sexual difference in investments at the beginning of the simulations. However, this dimorphism disappears as the simulations continue, resulting in essentially no dimorphism by the end.

Experiment: Dimorphic Minimum Investments

We can test the hypothesis in a more forgiving way by configuring the simulation differently. Namely, we can set a minimum — non-evolvable — difference in the investments that the sexes make. Since evolution cannot remove this difference in investments, a further difference in investments has a better chance of evolving.

We tested whether this occurs by forcing females to make a larger minimum parental investment than males. We then checked for a divergence in the evolution of total parental investments that both sexes make. Here, *tpi* will stand for total parental investment (as before), *mpi* will stand for *minimum parental investment* and *epi* will stand for *evolvable parental investment*. Thus, $tpi = mpi + epi$, and $mpi_f > mpi_m$ — that is, females have a greater minimum investment than males.

Results. Some of the results from this test can be seen in the graphs of the right column of Figure 5.18. In the first graph, $mpi_f = 10$ and $mpi_m = 0$. In the early period of this run set, both sexes evolve greater *tpi* and the difference remains roughly equal to the initial difference. However, as the simulation proceeds, tpi_f and tpi_m slowly draw towards the same amount. In the second graph, $mpi_f = 100$ and $mpi_m = 10$. Here, both sexes evolve the same further investment ($epi = 10$). Thus, the difference in *tpi* remains the same. Ultimately, then, greater investments by females do not evolve in either run set.

Figure 5.18f shows the reproductive success distributions for the run set of Figure 5.18d (again, for the last ~7000 cycles). We can see that a substantial dimorphism in reproductive success develops — the KL divergence here is 0.023 — with males exhibiting greater variance. This is due to the large difference in *tpi* between the sexes for this run set, and further confirms that greater variance in reproductive success will be shown for the sex that invests less.

The reason for the difference in reproductive success distributions for this latter run set can be discovered from the bottom row of Table 5.5. We can clearly see that a substantial dimorphism evolves, with males mating more frequently, and females choosing to eat more frequently. This is exactly what parental investment theory predicts — that is, that the sex that invests less will evolve to try mating more often. Of course, trying does not equate with succeeding — males (and females) must average 2 offspring in a stable population. Instead, the greater sexual activity of males leads some to greater success, and this in turn causes others to have less success, as we see in Figure 5.18f.

5.4.3 Desertion Hypothesis

The desertion hypothesis was born from Dawkins and Carlisle's criticism of Trivers' prior investment hypothesis (Dawkins and Carlisle, 1976). Dawkins and Carlisle noted that dimorphic investments may evolve when only one parent is needed to raise a viable offspring. In particular, if one sex has a chance to desert the offspring first, then it will. Dawkins and Carlisle cited

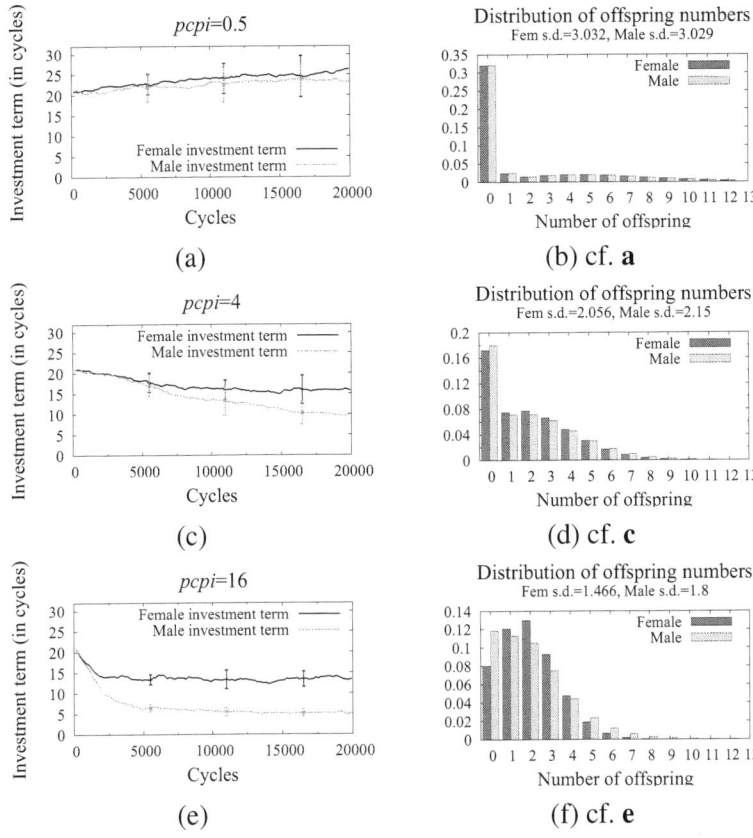

Figure 5.19: The evolved *eit* for males and females when (a) *pcpi* = 0.5, (c) *pcpi* = 4 and (e) *pcpi* = 16. Distributions of reproductive success by sex for the simulations in (b) **a**, (d) **c** and (f) **e**.

parental investment amongst fish as an example: in many species of fish, it is the male who looks after the offspring. They suggested that this is because females spawn their eggs first and males fertilize them after — by which time, of course, the female has the opportunity to leave. In contrast, male mammals fertilize female eggs internally, producing zygotes within the female. In those cases, the male clearly has the first opportunity to desert, potentially explaining why *maternal* care (which occurs after birth, of course) is predominant amongst mammals.

154   CHAPTER 5. EXPERIMENTS IN EVOLUTION

Experiment: Varying Per-cycle Parental Investments

To test this hypothesis, we allowed parents to invest for an evolvable period after birth (the evolvable investment term, or *eit*). For females, we set the minimum eit_f to 5 cycles; in contrast, males had no minimum period (other than 0, of course). The child required a minimum investment of 32 cycles — so if both parents invested for the same terms, they would each invest at least 16 cycles. If one parent quit investing before 16 cycles, the other parent was forced to compensate. Finally, we fixed the *per cycle parental investment* (or *pcpi*) as a parameter of each run set. In the simulations shown here, the *pcpi* takes on the values: 0.5, 4 and 16 health units per cycle.

Results. Figure 5.19 shows the results of the tests of the desertion hypothesis. When the *pcpi* is lowest, no dimorphic investments resulted (Figure 5.19a). In this case, relatively high periods of investment are needed from both parents: each tries to invest for ~25 cycles, which results in ~50 cycles of combined investment — well above the minimum 32 cycles of investment needed by the child. Thus, the female's minimum *eit* of 5 cycles becomes irrelevant. In the reproductive success distributions for this run set, shown in Figure 5.19b, we can see that no substantial sexual difference exists (the KL divergence is 4.8×10^{-5}). Furthermore, there is no sexually dimorphic behavior evident either (first row, Table 5.6).

For the run set in which *pcpi* sits at the higher level of 4 health units per cycle, the result is very different. Here, eit_f reaches an average of 15 cycles, while eit_m reaches an average of ~10 cycles. Since $eit_m < 16$, females must make up the remaining investment, so females invest for the greater of $eit_f \approx 15$ and $16 + (16 - eit_m) \approx 22$ — that is, about 22 cycles. It is interesting that the minimum eit_f of 5 cycles can have an effect here. In fact, the average standard deviations of eit_m and eit_f (not shown in the graphs) fall between 5 and 7 cycles, allowing the minimum eit_f to influence the evolution of investments.[5] Note that Figure 5.19d shows that dimorphism in reproductive success begins to develop in this run set.

Finally, in the $pcpi = 16$ run set shown in Figure 5.19e, eit_m reaches 5 cycles and eit_f reaches 15. That is, females come to invest for ~27 cycles. This establishes strong conditions for dimorphism — which indeed evolves, as can be seen clearly in Figure 5.19f and the bottom row of Table 5.6.

---

[5]The confidence intervals shown in the graphs only report the variance across the *run averages*, not the variance in the underlying populations.

CHAPTER 5. EXPERIMENTS IN EVOLUTION

	Eat	Mate
$pcpi = 0.5$	0.38% (59.1%)	-0.52% (29.6%)
$pcpi = 16$	7.5% (71.6%)	-7.2% (20.1%)

Table 5.6: Female minus male action rates for the desertion experiments, with female action rates in parentheses.

5.4.4 Paternal Uncertainty Hypothesis

The paternal uncertainty hypothesis is again due to Trivers (1972). He suggested that males are often uncertain about their parentage, particularly in species where females go through a gestation period. In contrast, uncertain female parentage is very unlikely. In that case, it may pay males to increase mating investment at the cost of parental investment: that is, spend less effort on an offspring, who likely is the child of another male, and instead spend more effort trying to mate again. There is some evidence in humans that paternal uncertainty has an effect on how parents and their families interact. For example, Daly and Wilson (1982) report that the mother's family will make comments about how similar the child looks to the father more frequently than reciprocal comments are made by the father's family. Further, Fox and Bruce (2001) report that paternal certitude affects how fathers take to their roles as fathers.

Experiment: Varying the Probability of Paternity

To test the paternal uncertainty hypothesis, we arranged for the probability of paternity, pp, to be a parameter of the simulation. In particular, females always invest in their own offspring; in contrast, males are chosen from the neighborhood to invest in their offspring according to pp. At one extreme, if $pp = 1$ for a simulation, the chosen male is *certainly* the father; at the other extreme, if $pp = 0$ for a simulation, the chosen male is *never* the father. We set pp to 101 equally spaced values between 0 and 1 inclusive. As for the prior investment experiments, the parental investments that both sexes make, tpi_f and tpi_m, are free to evolve.

Results. Figure 5.20a shows a scatter plot where each point represents the average tpi_m of the last 1000 cycles (of 7000 total) in a single run. The horizontal axis shows the setting of the pp parameter for each run, and the vertical axis indicates the investment amount. The result here is clear: the lower the probability of being the actual father, the less males invest in the

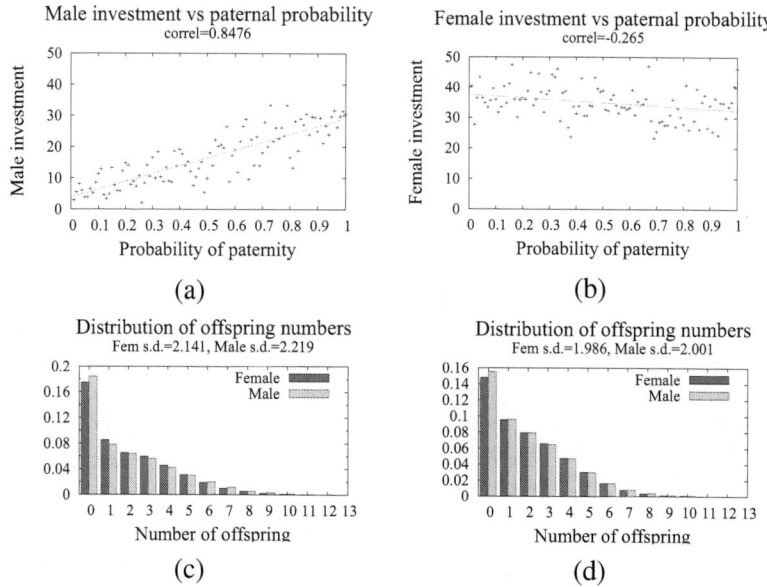

Figure 5.20: (a) Evolved investments made by males as a function of *pp*. (b) as per **a**, but for females. Distributions of reproductive success by sex for (c) $pp < 0.1$ and (d) $pp > 0.9$.

	Eat	Mate
$pp < 0.1$	3.0% (67.2%)	-3.1% (19.2%)
$pp > 0.9$	0.72% (66.9%)	-0.86% (19.4%)

Table 5.7: Female minus male action rates for the paternal uncertainty experiments; female action rates in parentheses.

offspring. Indeed, *pp* and tpi_m have the very large correlation coefficient of 0.848 ($t(100) = 15.81$, $p < 0.001$). This result provides strong support for the paternal uncertainty hypothesis.

We can also see how females evolve tpi_f for different *pp* from Figure 5.20b. As *pp* increases, and as males therefore invest an increasing amount, tpi_f falls away slightly. The negative correlation is not large ($r = -0.265$), but is statistically significant ($t(100) = -2.72$, $p < 0.004$). Thus, the more males invest, the more females take advantage by investing less.

To assess the level of dimorphism (in behavior and reproductive success) that evolves in these runs, we took the runs in which $pp < 0.1$ as one group and $pp > 0.9$ as another. The former should exhibit more dimor-

CHAPTER 5. EXPERIMENTS IN EVOLUTION

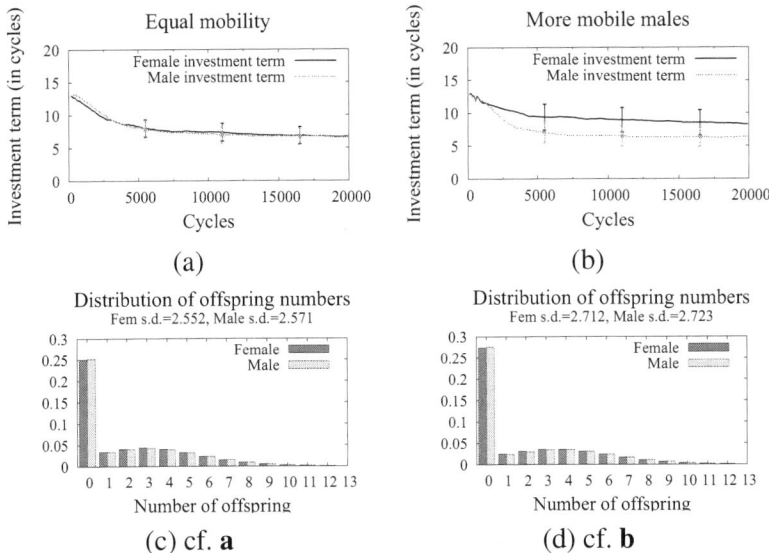

Figure 5.21: The evolved after-birth investment terms for both males and females for (a) no sex differences and (b) males as the more mobile sex. Distributions of reproductive success by sex for (c) no sex differences and (d) males as the more mobile sex.

phism, while the latter should exhibit less. Figure 5.20c and Figure 5.20d shows the reproductive success distributions for the last ~1200 cycles of runs with $pp < 0.1$ and $pp > 0.9$, respectively. We do find some very minor dimorphism in the $pp < 0.1$ runs (KL divergence of 0.0011) that is not evident in the $pp > 0.9$ runs (KL divergence of 0.00019). Similarly, we can see a minor behavioral dimorphism in Table 5.7 for the $pp < 0.1$ runs, that is reduced in the $pp > 0.9$ runs.

5.4.5 Association Hypothesis

The association hypothesis, or the pre-adaptation hypothesis, was suggested by Williams (1975) and points out that if only one sex remains in the vicinity of the offspring after birth — due to some pre-adaptation of that sex — then that sex has the opportunity to evolve parental care, while the other sex does not.

	Eat	Mate
Equal mobility	0.20% (68.0%)	-0.18% (23.8%)
More mobile males	1.1% (66.2%)	-1.1% (25.2%)

Table 5.8: Female minus male action rates for the association experiments; female action rates in parentheses.

Experiment: Resource-intensive Males

If females are pre-adapted by necessarily being in the vicinity of offspring at birth, there is nevertheless no a priori reason why males cannot remain nearby or return and invest themselves. Of course, males will find it *harder* to invest in offspring, meaning that males have to do more to invest at the same rate as females.

We tested the effects of association by having a non-evolvable 'Move' action that causes males to move about more actively. Specifically, males move about randomly in a 9x9 neighborhood with 0.6 probability each cycle, while females move about randomly in a 3x3 neighborhood with 0.2 probability each cycle. In addition, we set varying degrees of efficiency for parental investments, depending on their distance from the child. That is, the closer a parent is to a child, the more of its investment reaches the child. We used a simple linear inverse function of distance from the parent, where the distance is (crudely) measured as the minimum number of cells either horizontally or vertically.

Similar to experiments above, the investments were in the form of per-cycle investments after birth. Here, $pcpi = 8$ and agents were free to evolve the term for which they invest (the *eit*).

Results. Figures 5.21a and b show the results of two run sets, the first in which the 'Move' action is the same for both sexes, and the second in which the 'Move' action makes males more mobile. The graphs show the *eit* for both sexes. The first graph shows no dimorphism evolving — as we would expect. In contrast, the second graph shows that females — the sex that could invest more efficiently — evolved to invest for longer periods.

Surprisingly, the degree of behavioral dimorphism that evolves was very slight. The bottom row of Table 5.8 shows that a difference of only 1% in action rates evolved — in contrast to experiments in previous sections that showed differences of between 3% and 7%. Further, dimorphism in reproductive success (not shown) was not evident (there was a KL diver-

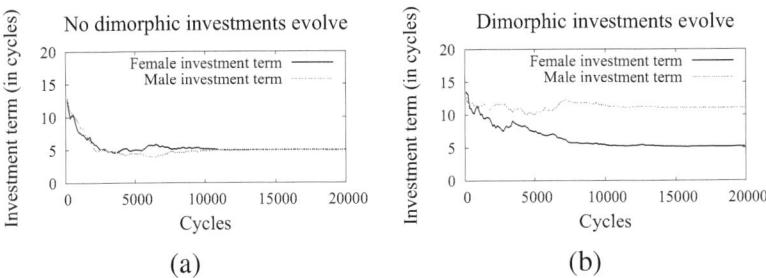

Figure 5.22: The evolution of *eit* by sex in single simulations with (a) no evolved dimorphic investments and (b) evolved dimorphic investments. Both simulations are taken from run set C of §5.4.6.

gence of 6.6×10^{-5} for the more mobile male run set).

5.4.6 Chance Dimorphism Hypothesis

Trivers (1972) suggested that the sexes may have differentiated in investment very early on due to positive selective pressure acting on gametes whose sizes fell in the tails of the normal curve. That is, smaller, more mobile gametes would be selected for since they can fertilize other cells more easily, while at the same time larger and immobile gametes would be selected for since they increase the probability of viable offspring. In contrast, those with intermediate sizes would not fare so well. Trivers didn't regard this process as recurring in new species, but rather as having occurred in some progenitor species, from which dimorphism was inherited.

However, it is also possible that sexual differences in parental investment can arise by chance. If an accidental difference in investments persisted for long enough, the sexes might adapt to the difference. This could then 'lock them in'. We would expect such events to be most likely amongst small populations, since the genetic variance within such populations will be smaller relative to the genetic variance between such populations.

Experiment: Degrees of Dimorphism by Population Size

To see if dimorphism can arise by chance, we ran simulations in which the sexes are initially identical across a range of population sizes and then measured for differences in investments ($|eit_f - eit_m|$) and behavior. We controlled the population sizes by regulating the food supply. To assess the degree of dimorphism, we use $dd = |eit_f - eit_m|/sd$, where sd is the pooled

Expt	Avg stable pop size	Degree of dimorphism (dd)	Distribution of dd
A	237	$\mu = 44.32$ $\sigma = 83.96$	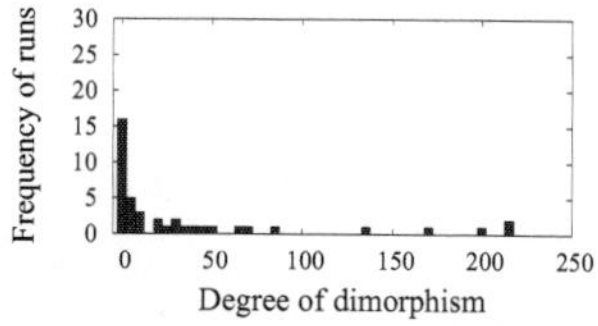
B	409	$\mu = 29.19$ $\sigma = 54.98$	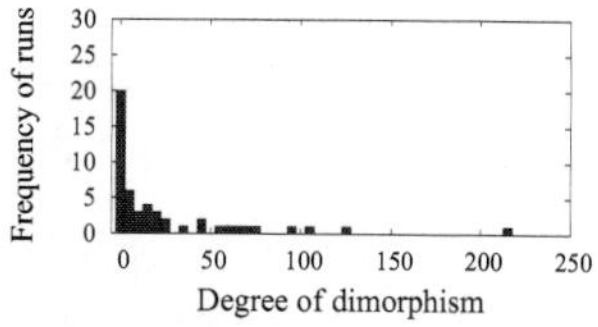
C	648	$\mu = 9.215$ $\sigma = 11.81$	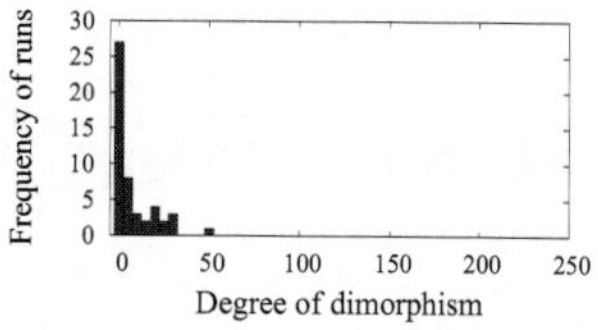
D	901	$\mu = 11.19$ $\sigma = 17.12$ $\mu* = 9.406$ $\sigma* = 11.78$	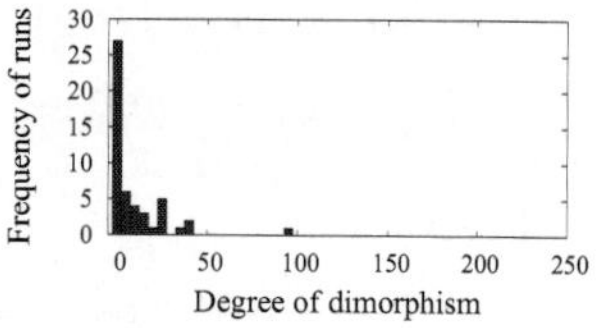
E	1198	$\mu = 8.208$ $\sigma = 10.50$	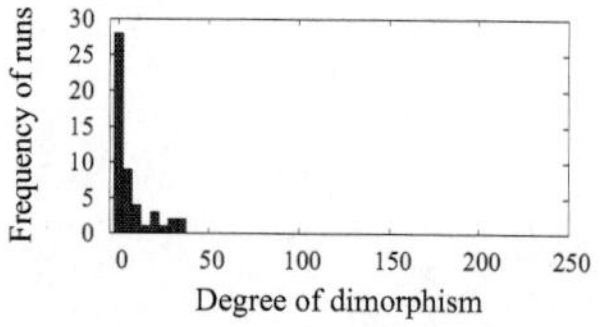

Table 5.9: Averages and standard deviations of degree of dimorphism (dd) for differing population sizes. Also, distribution of dd graphs (rightmost column); each bar in the graph spans 5 values of dd. (*) For the statistics for experiment D a significant outlier has been removed.

CHAPTER 5. EXPERIMENTS IN EVOLUTION

	Eat	Mate
No dimorphic investment	0.44% (78.8%)	-0.20% (18.8%)
Dimorphic investment	-2.7% (75.3%)	2.7% (21.5%)

Table 5.10: Female minus male action rates for the single chance dimorphism simulations of Figure 5.22. Female action rates in parentheses.

standard deviation of eit_f and eit_m within a run. I.e., dd is the number of standard deviations of difference between eit_f and eit_m.

Results. Table 5.9 summarizes the results of five run sets, each with different average population sizes. The table shows the mean dd for a run set with a given population size (along with the standard deviation). Dimorphism evolves quite regularly, with Table 5.10 also showing behavioral dimorphism from run set C. If we focus on those cases in Table 5.9 in which there are three standard deviations or more of difference (i.e., $dd \geq 3$), we see that dimorphism results in half or more of all cases. Further, there is an inverse correlation between the size of the population and both the average and variance of the dd that evolves. Experiment set D here defies this trend, but only with a minor deviation which itself is largely due to a single large outlier. The general trend of the dd distributions clearly confirms our expectation that chance dimorphism is far more likely to arise in small populations.

5.4.7 Summary

We have explored some of the main contending hypotheses on the origins of sexually dimorphic parental investment using our ALife simulations, and found support for those which provided clear rationales for the origin of divergence. The simulation results reinforce Dawkins and Carlisle's (1976) criticism of Trivers' prior investment hypothesis, failing to find evidence of such an origin. On the other hand, the desertion hypothesis and the paternal uncertainty hypothesis were easily modeled and readily demonstrated. While our results are also consistent with the association hypothesis, the level of dimorphic behavior and reproductive success in these experiments were lower. Finally, we had little difficulty in finding simulations that produced dimorphism by chance and confirmed that the smaller a population, the greater its chance of originating dimorphism. Aside from prior investment, in real populations any or all of these causes may be acting to bring about sexually dimorphic parental investments or in maintaining existing dimorphism.

5.5 The Evolution of Utility

The evolution of language and cognition has long been a focal point of research. Understanding them and applying insights about them are at the heart of much of evolutionary psychology. And there have been a fair number of attempts to model the evolution of cognitive structure, mostly in the form of evolving artificial neural networks for predicting and guiding behavior, for example, in Larry Yaeger's well-known simulation, Polyworld (Yaeger and Sporns, 2006, Yaeger et al., 2008). However, there has been, thus far, little or no serious effort put to evolving decision making as such.[6] Rational decision making, in the classic view (e.g., Ramsey, 1931), requires maximizing one's expected utility by, first of all, learning what one can about the consequences of possible actions and then assessing those actions by computing their probability-weighted implications for utility. In other words, rationality is the selfish utility maximization of Equation (2.5):

$$v(a) = \sum_j u(o_j) p(o_j|a)$$

Ignoring for the moment the implicit slur on utilitarians, we can agree that the study of evolving cognition ought to proceed in a broader context than the evolution of probabilistic predictive models. In particular, what has been missing is the evolution of utility and its combination with the evolution of predictive models in the evolution of decision making. Cognition may be modular, but that doesn't mean the evolution of one module can be well understood in isolation from the evolution of the rest.

5.5.1 Utility and Fitness

How do evolved utility and fitness relate to one another? The most intuitively pleasing answer is the simple Positive Association Thesis: that they relate directly and inexorably together. This is the not-very-hidden meaning, in fact, of the title of Jared Diamond's (1997) book *Why Is Sex Fun?*

---

[6]Paul Glimcher (2004) describes very interesting research showing that the *result* of evolution in our case includes neural processes in the parietal cortex that closely correlate with the probability of reward and also with the magnitude of reward. That we are beginning to find neural correlates of the ingredients of rational decision making, both probabilities and utilities, just reinforces the need for simulation studies of the evolution of the same.

CHAPTER 5. EXPERIMENTS IN EVOLUTION

How could it be otherwise, at least for those of us still around to enjoy the title? Unfun sex is unfit sex. Those activities with positive utility are those which a rational creature chooses to undertake; if these activities are systematically unfit, then these rational creatures will over time disappear. For them, extinction is rational.[7] The same applies to activities with negative utility that have positive fitness, for rational creatures by foregoing them will tend to present opportunities for competing individuals within their species to increase their relative fitness, and so, again, said rational creatures will be rationally selecting themselves out of the genetic pool. Thus, the question, "Why is suicide unpleasant?" has a similarly automatic answer: any who found it otherwise have long since departed. Conclusion: utility must be strongly, positively and unambiguously correlated with fitness, i.e., the Positive Association Thesis.

That is not the answer delivered by Samuelson and Swinkels (2006). They start their argument from a different point, asking the question, "What is the value of utility?" Their generic answer is consistent with recent work on the psychology of emotion and also **reinforcement learning** theory in AI, namely that utilities serve to direct short- and intermediate-range decision making towards ultimate purposes. In reinforcement learning, actions which have immediate rewards (positive or negative) contribute some "tax" on that reward to actions in the recent past which have helped to reach the state where the final act could be rewarded. In effect, the intermediate actions build up a history of rewards into stable utilities associated with those actions, with the result that the more useful actions will be chosen more frequently over time. Utilities, then, are a kind of currency being used to boost the choice of some actions and suppress that of others.

But, then, why would anyone need to have some actions boosted and others suppressed? If we were all perfectly rational actors, not only in the sense of maximizing our expected utilities, but also in the sense of having optimal expectations in the first place and, in particular, having perfect information-processing facilities for predicting future states, then we would always be in a position to maximize our expected fitness without any reference to utilities. The ultimate influence of any intermediate actions on our expected fitness would be no mystery to us, and there would be no evolutionary advantage to boosting or suppressing our tendencies to act in any

---

[7]In much of this discussion "activity" and "action" will be used as a kind of shorthand for an action within a context. One and the same basic act may accrue quite different utilities under different circumstances, of course; but the shorthand can be justified as a simplifying maneuver, and by the fact that actions can always be explicitly qualified by their circumstances.

way. The *only* adaptive role utilities can have is to compensate for shortcomings in predicting future fitness outcomes. That is not a minor role, of course. The fitness outcomes of most actions are very long-range outcomes, and no one is in a position to calculate them with any exactness. If we then accept that utilities are strictly adaptive (perhaps just for the sake of argument), then positive utilities must be associated with actions that need boosting, that is, which lead to outcomes that are systematically undervalued by the agent's fitness calculations, and negative utilities are associated with those actions that are systematically overvalued.

So far, so good. Positive utilities are associated with some amount of underestimated fitness, the gap between expected fitness as measured by the agent (or, simply, "expected fitness") and as measured by an objective probability (an enduring frequency in an environment of evolutionary adaptation, which we can call simply, "actual fitness", bearing in mind that this, too, is an expectation). Negative utilities likewise measure the extent to which the fitness of actions are being overestimated by the agent. We can call this the Gap Theory of Utility.

This leaves open the question of whether there is any systematic relation between utilities and fitness. Samuelson and Swinkels (2006) argue that there can be no systematic relation, that the two must be uncorrelated.

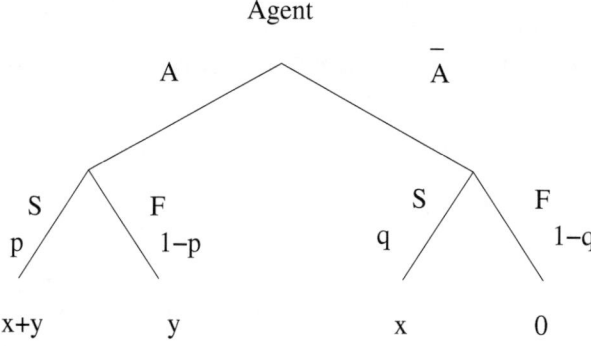

Figure 5.23: A generic decision-making scenario with utilities.

Samuelson and Swinkels' Argument

Under various assumptions, the plausibility and generality of which we will address below, Samuelson and Swinkels (2006) actually *prove* the Gap Theory. We reproduce the main points in their argument here, but without

CHAPTER 5. EXPERIMENTS IN EVOLUTION

presenting their formal proof.[8] The situation they consider is represented in Figure 5.23. This is a highly simplified decision-making scenario. The agent is contemplating taking the action A or the alternative action $\overline{A}$. The consequences of interest are "success" (S) or "failure" (F), which are actually successful reproduction or failure to reproduce, respectively, since the issue here is one of utilities' role in determining biological fitness. y is the utility associated with taking action A and x is the utility associated with the successful outcome. The utility associated with the action $\overline{A}$ is fixed at 0 for simplicity, which is an arbitrary decision that helps to fix the utility scale. Another simplifying decision that finalizes that scale is to set $x = 1$. They require the agent to have reasoning and inferential capacities with which to make predictions about the success or failure of its actions, which they call the function $\phi(\cdot)$. This is a function from scalar "signals" to probabilities, where the signals reflect the percepts of the agent that are relevant to estimating p and q and are designated s_p and s_q respectively. ϕ is assumed to be continuous and strictly increasing, a "consistency" assumption the authors consider likely for evolved agents.

Let $\phi(s_p) = p', \phi(s_q) = q'$ and $\Delta = p - q$. The agent's estimate of Δ is $\delta = p' - q' = \phi(s_p) - \phi(s_q)$. Perfect rationality under perfect information dictates the optimal decision rule: do A if and only if $\Delta > 0$. That is to say, this is the rule we would expect evolution — reproduction of the fittest — to impose if it could. Of course, agents can't implement this rule, but instead attempt to approximate it by maximizing their expected utility, which gives us the suboptimal decision rule to choose A whenever:

$$
\begin{aligned}
p'(x+y) + (1-p')y &> xq' \\
p'(1+y) + (1-p')y &> q' \\
p' + p'y + y - p'y &> q' \\
p' + y &> q' \\
p' - q' + y &> 0 \\
\delta + y &> 0
\end{aligned}
$$

This practical decision rule is equivalent to the optimal decision rule when and only when $\Delta = \delta + y$, i.e., $y = \Delta - \delta$, which is just to say the utility y exactly covers the gap between the predicted probabilities and the actual

---

[8]To be sure, Samuelson and Swinkels (2006) are not explicit in their treatment of cognitive architecture and leave a lot of holes in their description, which we fill in.

probabilities. While there are considerable further details to Samuelson and Swinkels' discussion, this is the gist of their argument.

It is perhaps worth noting in passing that Samuelson and Swinkels' model of decision making is consistent with the Bayesian theory of rational decision making. Although in some sense the question is raised by it of a tradeoff between maximizing utility and optimizing predictions of fitness, the fitness associated with a successful reproduction is itself given a fixed utility of 1. On the scale adopted, 1 must actually be a very high utility, since otherwise, as Diamond suggests, rational maximizers of utility would become scarce.

Under the assumptions that Samuelson and Swinkels adopt, there is no question that they are correct and Gap Theory is proved. Those assumptions, however, are very strong, far stronger indeed than the authors acknowledge, for they imply that their argument is valid for almost all evolutionary situations. We think, on the contrary, that it is valid only for extremely simple evolved agents and not for agents having any substantial cognitive capacity. This doesn't mean that Gap Theory is wrong, but that it remains unproven. Very likely its status will remain open until evolutionary experiments resolve the issue, experiments we intend to make.

The main doubts we have concern the function ϕ. Its inputs, s_p and s_q, boil down all of the observational evidence available to the agent about reproductive success to a pair of scalars. It is certainly not the case that perceptual cues of the reproductive success of actions are unidimensional. For example, potential mates are assessed upon multiple criteria, such as health, quality of the immune system, availability, etc. It may nevertheless be possible to summarize them all in a single scalar; it certainly makes the mathematical model simpler and is required for Samuelson and Swinkels' proof. Regardless, it is clear that the ϕ corresponding to actual animal brains are not univariate functions. The best estimates of the probability of reproductive success given an action A is in most cases likely to be a function not just of a signal s_p summarizing perceptual input relevant to predicting its success, but also jointly a function of the agent's history with similar signals and similar actions; cognitive capacity in our world comes with an ability to *learn*. In addition to that, cognitive capacity in our world often comes with an ability to introspect, including an ability to appreciate and respond to the utilities one experiences. The authors claim, without argument, that it is unlikely that evolution would allow agents to make inferences based upon their emotional states. But this is contrary to the fact that many agents have cognitive access to their emotional states and contrary to the fact that

CHAPTER 5. EXPERIMENTS IN EVOLUTION

evolution takes advantage of whatever is available to it to improve fitness. In short, the form of ϕ ought to be:

$$\phi(\vec{s}_p, \vec{p}, \vec{y}, \vec{x}) = p'$$

where each argument corresponds to the *sequence* of relevantly similar decisions in the history of the agent. This relaxation of the restrictions upon ϕ makes the model more realistic and generally applicable and also invalidates their proof of the Gap Theory, which relies upon those restrictions.

Gap Theory and Association

We should hope that a theory of the relation between utility and fitness would tell us something definite about their association, positive, negative or neutral. It seems, however, that the Gap Theory is actually compatible with any kind of association between utilities and fitness — i.e., it tells us nothing about it. Suppose, for example, that the actions available to an agent are normally distributed for fitness with a mean of zero. Suppose also that for each action its fitness as estimated by the agent is normally distributed with a mean of its actual fitness. In that case, the gaps that need to be covered by utilities will be normally distributed with a mean of zero at every point on the fitness scale, with the consequence that there will be no correlation between utilities and fitness. These may seem like natural assumptions, but we know of no serious argument on their behalf. And it seems there are alternative assumptions that are at least as natural.

For example, the best prediction of future states in cases of total ignorance is often one that maximizes entropy, that is, one that grants equal probability to all recognized possible outcomes. For example, if you know nothing about football, you will only maintain calibrated predictions by giving teams a 50-50 chance of winning (ignoring draws). Our ignorance of future states may be due either to a lack of relevant factual information, or to an inability to model state transitions, or to our inability to process available facts to make effective predictions. In other words, to the extent that our predictive information processing is limited either by input information, modeling ability, or simply by computational abilities, it should tend to result in estimated outcomes that fail fully to discriminate between alternatives. If these computational abilities for predicting future states have evolved even partially as general-purpose processing facilities, then, when recruited for the purpose of estimating future fitness, it may readily turn out that this "conservativeness" manifests itself as a systematic *underesti-*

mation of the variance of the fitness of actions — in other words, the fitness value of arbitrary actions will tend to be estimated as near zero.

The evolution of more and more sophisticated intellectual decision-making capacities may be seen as a long-term evolution of the capacity to correctly estimate fitnesses of actions that are distinct from zero. But in that case, so long as we do not adopt the strange conceit that we are the epitome of intellectual evolution, we must recognize that much of our fitness computations will be constrained by ignorance, modeling limitations and computational incapacities — in other words, that our estimates of fitness will be systematically biased towards zero and away from extremes. From this assumption a positive correlation between fitness and utility immediately follows. Systematic underestimation of variance (combined with no systematic bias in estimation) implies that actions with positively estimated fitness will be underestimated in general, while negatively estimated actions will be overestimated in general, so the gaps postulated by Gap Theory will be positive for the more fit actions and negative for unfit actions. And this result is entirely in keeping with the central points promoted by Samuelson and Swinkels (2006), which just is Gap Theory.[9]

None of this argument shows that no contrary argument is possible. In fact, it is not really harder to put together an equal and opposite argument. If fit actions are systematically overestimated and unfit actions systematically underestimated, then the correlation obtaining between fitness and utility may be just as strong, but in the opposite direction. Such overestimation of the magnitude of value for fit and unfit actions is, for example, suggested by the psychological evidence of systematic overconfidence in probability estimation. Since we have a natural tendency to overestimate the probability of outcomes we are inclined to believe will occur, this will likely manifest itself in overestimating the magnitude of impacts on fitness of intermediate actions. So we can argue for positive association and for negative association. Presumably, then the arguments can be combined in arbitrary ways, leading to any kind of association between utility and fitness. Thus, there seems to be no a priori reason to think that overestimation of fit actions might not be combined with overestimation of unfit actions, or again underestimation of both, when again no correlation between fitness and utility could be expected to be found. Positive and negative associations between

---

[9]Samuelson also writes, "In general, evolution should make us fear not simply things that are bad for us, but rather things whose danger we may underestimate *without* discovering our error before they kill us" (Robson and Samuelson, 2008). These would be cases of negative utility complementing an underestimated negative fitness magnitude.

fitness and utility may be found only piecewise along the utility scale. Or again, perhaps the association is radically non-linear. In short, it seems that Gap Theory is consistent with any kind of relation at all between utility and fitness. It fails to give us an answer; what it does is provide us with a language in which to frame and examine all possible answers.

5.5.2 Design of an Experiment

The Positive Association Thesis of Diamond comes with no theory: Diamond does not present an argument in favor of it of the kind we have just given. What it does come with is a great deal of evidence in its favor. Sex is fun; so is eating nutritious food; so is playing a sport. On the other hand, getting injured is unfun; likewise eating most harmful food; likewise going hungry or thirsty. With the notable exception of overindulging in fatty and salty food, which evolutionary psychologists attempt to explain away as maladaptations based upon an EEA in which such behaviors were adaptive, Positive Association appears to receive overwhelming inductive support. We would like to know whether the Thesis is correct in general or under some restrictions and, if so, what. We would also like to know where and when Gap Theory is correct, beyond the trivial domains in which it has been proven to be correct.

Gap Theory, while having no implications about the Positive Association Thesis or its alternatives, does have some implications about how one should go about experimentally investigating the relation between utility and fitness. In particular, the Gap Theory itself can be tested by recording evolved utilities and the difference between actual and estimated fitness values associated with them in some particular evolutionary environment. They should fit like hand and glove. In order to test Gap Theory and the various association theses, we have designed a simulation in which cognitive capacity for predicting future states, including estimating the fitness of various behaviors, can co-evolve with utilities. The brains of our agents, our implementation of ϕ, are composed of a set of **naive Bayes** predictor models, allowing multimodal observations of the environment and the self to be used to predict future states (Korb and Nicholson, 2010, Chapter 7). The Bayesian models both evolve over the generations and themselves improve through the life spans of the agents. The utility functions only change over generations. Since the agents employ a version of the Bayesian ideal for rational decision making, selecting as their next actions those which maximize expected utility by combining their predictions and utility functions, evolution selects for fitter decision making, forcing both utilities and cognitive capacities to improve.

Although we have designed this simulation for evolving utility, and have implemented a preliminary version, at the time of this writing we have not conducted the proposed experiments and so cannot settle the questions we have raised. *Excepting* sex, we don't know why anything is fun.

In addition to answering that kind of question, we also plan to use the model to explore the evolution of *utilitarians*, agents who cater to the utilities of others in their decision making. We expect to find utilitarians evolving in situations with kin or group selection pressure that favor the emergence of altruism.

5.6 Conclusion

Evolutionary artificial life simulation is a tool of potentially great power in application to a huge variety of problems and is already in wide-spread use in the social sciences and epidemiology. While applying it ourselves to questions about the evolution of altruism and ethical behavior, we found ourselves wondering more and more about the mechanisms of evolution themselves. In investigating the debates about evolutionary action, we were struck by the inconclusiveness of the arguments and, more especially, of the available empirical evidence. ALife simulation here, as elsewhere, promises to supply genuine and important evidence for resolving many of these debates. Our work here represents only the very early beginning of such a theoretical application of a very practical method.

Chapter 6

Experiments in Ethics

6.1 Introduction

Having shown that ALife simulation affords new opportunities to investigate and test issues in evolution theory, we now look at its application to some of the problems in applied ethics, a kind of application that was our original motivation for pursuing this kind of simulation. The basic idea is that a better understanding of the evolutionary circumstances in which a kind of ethical behavior comes about can inform our understanding of that behavior, even if it can't, as the naturalistic fallacy would have it, simply issue a verdict on the behavior's ethical standing. We would like to know, ideally: when a target behavior can and cannot evolve; when it can, what gives it adaptive value, and when it can't, what prevents it from having adaptive value; what are the consequences of its presence or absence, and, in particular, whether its presence confers a utilitarian advantage on the population having it. Eventually we might be able to draw a fairly detailed map of an ethical behavior over the evolutionary fitness landscape, with annotations indicating how the fitness of the behavior relates to its utilitarian ethical value. In this chapter we report what are, again, the results of early and tentative explorations in this landscape, with just a few — rather contentious — kinds of behavior the targets of our investigations.

Utilitarianism requires too much calculation!

This is the claim of many critics of utilitarianism. It would be a pointed criticism if ethics and ethical behavior had to wait upon the outcome of exact and accurate utilitarian computations. But utilitarianism is thoroughly pragmatic: if, for example, the choice is between squeezing out an additional digit of accuracy in predicting the location of some prize by continuing computation, but at the cost of being too late, or of making an immediate best guess with any larger chance of finding the prize, utilitarianism opts for the latter. More generally, utilitarianism obliges one to be as accurate as one can, and not more so. Utilitarianism is based upon people making their *best efforts* — as, indeed, is the broader concept of ethical responsibility itself. While there is some evidence that some of our inferential capacities are Bayesian (e.g., Glimcher, 2004), the evolution of the "fast and frugal" heuristics of Gigerenzer, Todd and others (e.g., Gigerenzer and Todd, 1999) is perfectly compatible with the rationality of Bayesianism and the ethicality of utilitarianism. Where our evolved best efforts are simple shortcuts, Bayesians and utilitarians are free to take the shorter paths along with everyone else. Still, there is no need to walk them blindly, and both Bayesianism and utilitarianism offer tools with which to measure the length and ultimate adequacy of those paths. If they turn out to be shortcuts to nowhere useful, we should also like to know about that.

In this chapter, then, we attempt to further develop the instrument of evolutionary simulations of ethical behavior and their utilitarian analysis, combining ethical and evolutionary investigation. We believe looking at them in combination is far more rewarding than looking at them in isolation, as is clearly demonstrated in recent research on the evolution of cooperation. Charles Darwin was already the first person to undertake such combined investigations, but he was in no position to take advantage of any relative of his Charles Babbage's Analytical Engine in doing so. We are.

6.2 Cooperation

Before diving into our own simulations we shall review some of the more important work simulating the evolution of cooperation and altruism. Evolutionary research on cooperation has been dominated by the iterated prisoner's dilemma ever since Axelrod conceived of running a computer tournament for testing strategies (see Chapter 4). This dominance has been particularly noticeable in the simulation community, where the simple numerical model underlying the IPD is easy to implement and study in code. But this alone does not explain the game's continuing popularity, for there are 77 alternative two-player games that are just as simple and easy to implement, as well as potentially relevant to the study of cooperation.[1]

---

[1] A taxonomy of 2x2 matrix games is given by Fraser and Kilgour (1986).

CHAPTER 6. EXPERIMENTS IN ETHICS

Interest in the prisoner's dilemma is fueled by what appears to be a paradox: while defection by both players is a Nash equilibrium when the game is played exactly once, merely iterating the game turns up successful long-term strategies that depend on violating this Nash equilibrium. Of course, there is no paradox: there is more to the story than mere iteration — learning plays a critical role as does uncertainty. But the real attraction of the IPD seems to have been a kind of fluke: the naming of the actions *cooperation* and *defection* by the game's originators, Dresher and Flood, for it is the idea of evolving cooperation that is compelling.

'Cooperation' is an exaggerated name for the act of staying silent in the one-shot prisoner's dilemma. Cooperation means *coordinated* activity aimed at achieving a common goal and, so, minimally some kind of communication. In the one-shot prisoner's dilemma there is no communication nor any shared goal, let alone an opportunity to coordinate to achieve such a goal. The iterated version provides some model for coordination, with, for example, the initial act of silence in tit-for-tat (TFT) potentially indicating a willingness to cooperate. But for the interpretation of TFT as a model of cooperation to make sense a fairly rich background context must be assumed: playing the IPD allows for signalling, but the shared goal must be provided. In the right environment, TFT will evolve goals or no goals, since in those environments selfishly accumulating maximal reward is *a* goal. Nowak and Sigmund (1993) further demonstrated the sensitivity of TFT to its environment by the finding that Pavlov wins under a wider variety of circumstances than TFT.

6.2.1 Cultural Evolution and the Stag Hunt

Many simulations that explore the evolution of cooperation and altruism employ explicit models of *cultural* evolution (Skryms, 2004, Boyd et al., 2003). Cultural evolution involves the transmission of traits, with variation and selection, among members of a population during their lifetimes. Cultural evolution models are often tantamount to certain simplified forms of biological evolution models: namely, those that lack a distinction between genotype and phenotype, and perhaps also lacking recombination. Skryms (2004), for instance, has claimed that conclusions about cultural evolution models can apply equally well to biological evolution models, although others such as Tennant (1999) have disputed this, given the lack of a distinction between genotype and phenotype. Given the generic similarity between them and their reliance on similar replicator dynamics, we should expect that cultural evolution models can sometimes be informative about biological evolution.

Skyrms has made extensive use of cultural evolution to explore the implications of the **Stag Hunt** game (Skryms, 2004, Skryms and Pemantle, 2000, Alexander and Skyrms, 1999). The two players have the choice of either cooperating to hunt the stag, with a large reward that is hard to realize (and which can be had only if they cooperate) or one or both can choose to hunt the more easily caught hare individually, i.e., defect. An example payoff matrix for the stag hunt is shown in Table 6.1. Whereas the prisoner's dilemma contains just one Nash equilibrium (where both agents defect) the stag hunt contains two: both cooperate and both defect. Skyrms argues that while this game has received much less attention than the prisoner's dilemma, it deserves more, particularly in the study of cooperation.

		Bob	
		Cooperate	Defect
Alice	Cooperate	(2, 2)	(0, 1)
	Defect	(1, 0)	(1, 1)

Payoffs: (Alice, Bob)

Table 6.1: The stag hunt payoff matrix. Rows represent Alice's choices, columns Bob's choices.

Skyrms has collaborated with Robin Pemantle in examining how agents evolve to play the stag hunt when they are capable of choosing their interaction partners (Skyrms and Pemantle, 2000). In their simulations, agents at first visit other agents uniformly randomly. During each visit, both agents play the stag hunt and receive payoffs, which in turn affect how likely they are to visit each other in the future. Stag hunters learn to visit other stag hunters and, more interestingly, hare hunters learn to visit other hare hunters. In addition, there is a fixed population-wide probability that agents switch strategies (i.e., hunt stag or hunt hare) to that of the currently best performing strategy in the population. When this switching probability is relatively high (0.1), just 22% of the simulations converge to stag hunting. When low (0.01), however, stag hunting evolves in 71% of all cases.

6.3 Altruism

Simulations of the evolution of altruism extend back to the 1970s. Levin and Kilmer (1974) created simulations that explored the potential for altruism to evolve under group selection. Each simulation contained ten demes of equal size, containing from 10 to 50 individuals. Each individual consisted of a single diploid locus with two possible alleles, altruistic (A) or egoistic (E), giving three types of individual: AA, AE and EE. The selection coefficients (i.e., individual fitness effects) for these three genotypes varied according to the experiment, but typically the AA genotype would cause a loss in individual fitness relative to the other two genotypes. Thus, the altruistic allele was individually recessive. In addition, demes had a probability of survival that was a direct function of the frequency of altruistic alleles within the deme; demes that died out were replaced by a random sample of copies from the population at large. Finally, individuals could migrate between demes, which Levin and Kilmer implemented by randomly selecting two individuals from two separate demes, ejecting one from the simulation and replacing it with a copy of the other.

While Levin and Kilmer easily found conditions under which group selection supported altruistic alleles, they believed these conditions were far too demanding to be met in the real world. Despite this (and after noting their desire to play devil's advocate) they argued in favor of group selection as a potential factor in natural evolution. Their main concern was that the most important evolutionary role of group selection may be in its synergistic relationship with kin selection, which is also something we have emphasized in §5.1.

More recent simulations have explored the role of kin selection specifically. Taking inspiration from the work of Nowak and May (1992), den Dulk and Brinkers (2000) explored how kin selection, supported by spatially viscous populations, can promote altruism in the prisoner's dilemma. Their simulation consists of a two dimensional grid in which each cell is occupied by an agent that plays either always-cooperate or always-defect against its eight neighbors. After each agent has played all its neighbors, they are replaced by the next generation. A cell is populated by replicating the most successful agent in its neighborhood. Running this simulation results in the population evolving to play always-cooperate across most ($> 95\%$) of the board.[2] It is kin selection that promotes always-cooperate:

---

[2] Amusingly, the remaining defectors establish patterns familiar from the Game of Life, such as gliders and blinking.

when den Dulk and Brinkers turned off kin selection, by placing offspring at random locations, always-cooperate quickly disappeared.

For an act to be altruistic outside of a biological context, it must harm the actor *in units of utility* and benefit the utility of others (in contrast to biological altruism, in which we substitute units of individual fitness for utility). den Dulk and Brinkers called the agents that play always-cooperate *altruistic*. On a straightforward interpretation of the elements of the simulation, this is entirely appropriate and accords with both definitions of altruism. In particular, if the payoffs to agents from playing the prisoner's dilemma are treated as utilities, then any agent who plays always-cooperate sacrifices its own utility (whatever the opponent's strategy) while boosting the utilities of the agents against whom it plays. Of course, this behavior is also biologically altruistic, since decreases in payoff reduce the agent's opportunity to reproduce. And as is often the case elsewhere, the biological altruism in den Dulk and Brinkers's simulation is supported by kin selection.

Boyd et al. (2003) added another piece to the puzzle of how altruism can evolve and be sustained. Building on an earlier analytical model by two of the authors (Boyd and Richerson, 1992), they developed a simulation based on cultural evolution and group selection that examines how the option of punishment may support altruism. In this simulation, agents repeatedly play a two stage game. The first stage is a form of n-person prisoner's dilemma: with a certain probability agents can choose to cooperate, which is a small sacrifice that benefits the group's chance of survival, or defect. The second stage permits those that cooperated to punish defectors, at some cost to the punisher. Thus, the simulation allows three strategies: contribute to the group without punishing defectors; contribute and punish defectors; or defect.[3]

Typically, group selection models such as these require small deme sizes in order to support altruism. When Boyd et al. ran their simulations without the contribute-and-punish strategy, as expected altruism disappeared at larger deme sizes. However, by introducing punishment, the groups continued to support altruism at much larger group sizes. Key to this result was the diminishing cost of punishment as defection rates fell. That allowed even weak group selection to support altruistic punishment, which in turn

---

[3]While Boyd et al. describe their simulation in the words of cultural evolution, there are significant similarities to the evolutionary dynamics of the earlier simulation by Levin and Kilmer, including the mechanism for migration, deme size equality and the individual and group fitness effects for cooperation (altruism) and defection (egoism).

CHAPTER 6. EXPERIMENTS IN ETHICS

supported altruism in the form of group contributions.

6.3.1 Food Sharing

Scogings and Hawick (2008) have produced another group selection model showing the evolution of altruism in the form of food sharing. Their simulation is based on a predator-prey model. The agents are governed by an ordered set of simple rules that remain fixed for the simulation. Predators that have enough health try to mate, otherwise they attempt to eat prey; if neither is possible they move around, either towards other predators (which increases the chance of breeding later), towards prey or at random. Prey have analogous behavior, although they also aim to avoid overcrowding and, of course, predators. In typical runs of this basic simulation, spatial patterns emerge in which spirals of predators chase prey.

Predators are the main point of interest since they are capable of altruism — prey are essentially just mobile food. Predators can exhibit altruism by helping less fortunate predator neighbors. They do this by finding a neighbor who has a lower quantity of health than they do, then donating as much of their own health as would make them equal.

	Abundant Grass	Scarce Grass
All Altruistic	predator modest increase prey substantial increase	
All Selfish		predators die out
50-50 Altruistic-Selfish	altruistic predators quickly die out	altruistic predators eventually triumph

Table 6.2: Altruism via food sharing.

To explore the evolutionary value of altruism, Scogings and Hawick ran two groups of three simulations (summarized in Table 6.2). The first group contained abundant grass (i.e., food for prey) and the second scarce grass. Furthermore, in each group, one simulation forced all agents to be altruistic, another forced them all to be selfish, while in a third, sharing behavior was allowed to evolve after an initial 50/50 split between egoists and altruists. When grass was abundant, altruism increased predator numbers modestly while substantially increasing prey numbers. That is, predators came to conserve their food source. If altruism was left to evolve, however, it disappeared quickly. By contrast, when grass was scarce, selfish predators died out: indeed, simulations of pure egoists failed to last beyond a few thousand cycles. Under such conditions, altruism became much more valuable, with simulations of pure altruists exhibiting strong stability.

If altruism is left to evolve when grass is scarce, we might still expect selfish behavior to evolve, since selfish individuals benefit as much from altruistic neighbors as do altruists. That, of course, ignores inclusive fitness supporting altruism, so perhaps we should expect some form of polymorphism to develop. In the initial stages of the simulation, selfish predators indeed performed better than altruists, however, as the simulation progressed, altruists eventually emerged triumphant. Clearly, there is some tipping point for grass abundance that determines the outcome between altruists and egoists.

Scogings and Hawick suggest that the driving force for the altruism in their simulation is "true group selection and not kin selection." While the predators in their simulation certainly form clusters that can be called groups (although with fairly porous boundaries), it seems that kin selection must also be operating within these groups to strengthen within-group relatedness. This could, of course, be tested by turning off kin selection as we have done in the previous chapter and as done by den Dulk and Brinkers (2000). We expect that such an experiment, while preserving geographical groupings, would cause altruism to collapse.

6.3.2 Altruistic Suicide

In §5.3 we presented our simulation of an evolutionarily stable suicide. In earlier iterations of these simulations, we also explored the ethical effects of suicide on the world of the agents, by attaching utilities to the actions of agents. We ran the same basic simulation with suicide as one of the possible actions, using fixed utilities for the success and failure of each action, given in Table 6.3.

	SUCCESS	FAILURE
WALK	5	0
TURN	1	–
EAT	10	0
MATE	25	0
SUICIDE	0	–

Table 6.3: The fixed utilities associated with the outcomes of actions used to obtain the suicide experiments shown in Figure 6.1. A dash indicates that the outcome for the action in that row is not possible.

CHAPTER 6. EXPERIMENTS IN ETHICS

Calculating the cumulative utilitarian effects of a specific act, e.g., a specific suicide, is straightforward enough. At the point in our simulation where the suicide occurs, we can fork the simulation, with one process containing the suicide and the other excluding it, and then compare the consequences. However, we are not interested in specific acts, but in the consequences of *kinds* of acts, and replicating the forking process for *every* act of suicide is, of course, impractical. Instead, we run two sets of simulations. In the first set, suicide is possible; in the second, it is disabled. This allows us to compare the ethical impact of suicide across the two worlds. If we can identify conditions in which the world with suicide shows higher utilitarian value, this suggests that under those conditions suicide can be the more ethical choice.

Recall that in the suicide simulations, the food supply is cyclical. Figure 6.1a shows the cumulative total utilities for two such sets of simulations.[4] The ripples show the deleterious effect of periodic food shortages on the utilities of the agents. Whereas the total utility in the simulations with suicide is initially lower, the situation soon reverses, with the cumulative total utilities of the two simulation types subsequently diverging. Figures 6.1b and 6.1c show that the average utility with suicide is greater at every moment, hence the lower total utility in the early cycles of the suicide simulations is entirely due to smaller population size. Despite smaller early populations, it is not long before the ethical virtue of helping others to survive droughts asserts itself in a higher cumulative utility. We should recall what is sometimes forgotten in discussions of utilitarian ethics: the ethical value of an action depends on not just the immediate but also the *future* utilities of agents. Here, the cumulative graphs of both average and total utilities settle into a roughly linear form only after about 400 cycles. After 800 cycles, total utility tells the ethical story: in this world (that is, under the circumstances of this simulation), suicide is often the ethical option.

The Altruism of Suicide

Whether an act is altruistic in ethical terms, that is, comes at a net cost to an individual while benefitting the utility of others, is often difficult to determine. In the case of suicide here, it is straightforward. In these experiments, agents always derive positive utilities throughout their lives. Suicide prevents an agent receiving further positive utility, so it comes at a net cost

---

[4]They reflect 44 simulations with suicide and 31 simulations without suicide.

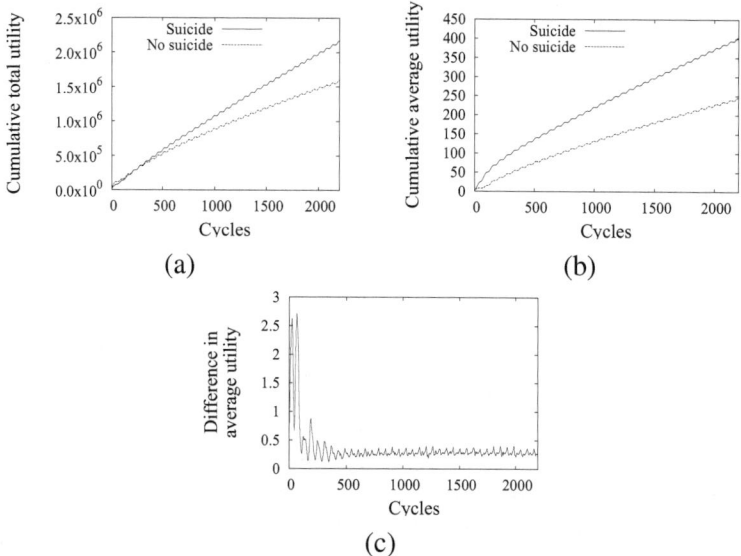

Figure 6.1: In simulations with and without suicide: (a) cumulative *total* utilities; (b) *average* utilities; (c) difference in average utilities.

to the agent. The question of whether suicide is altruistic, then, reduces to whether other agents benefit from suicide. Figure 6.1a shows that the total utility of agents is higher in the simulations with suicide. Obviously, the total utility of agents can only be calculated for agents that have not committed suicide — thus, it is the utility of *other* (non-suiciding) agents. Since suicide increases the total utility of other agents, suicide is indeed altruistic.

Some might consider an act to be altruistic only if it increases the *average* utilities of other agents rather than the total utility over the population. Figure 6.1c shows that also by this definition we find that suicide is altruistic.

Thus, our suicide simulations show a convergence of adaptive value for suicide with its altruistic nature, both ethically and in biological fitness terms. Of course, the ethical results above are sensitive to the utility structure we have imposed on the simulation and to various details in the design of the simulation. For example, we did not explore the possibility that suicide of an agent might have negative utility for other agents in the simulation, or, more simply, that suicide might have immediate negative utility for the actor itself. But, on the other side, we did not explore the possibility that starving to death might have a high cost in negative utilities. If we add to the utility cost of suicide, it is not hard to introduce compensating benefits, and vice versa. What we have shown is that where suicide reduces a substantial burden on a community under high stress, then both the biological and ethical consequences are likely to be beneficial.

6.4 Rape and Sexually Dimorphic Behavior

We move now from modeling typically ethical behavior to modeling clearly *unethical* behavior. In the previous chapter, we explored several theories from evolutionary biology that explain how the sexes may evolve to invest differently in offspring. Here we examine the consequences of this dimorphic investment for the evolution of rape and, by way of contrast, consensual mating. We also consider what factors may change the ethical impact of rape.

In these experiments 'rape' refers to any non-consensual act of mating, perpetrated by one agent but not agreed to by the other, its victim. Furthermore, it is a behavior that may harm the victim and may lead to conception. Obviously, we believe that there is an analogy with rape in the natural world, however, we focus here on how rape may evolve as a **reproductive strategy**, particularly in the face of punishment costs and consensual mating opportunities.

Many social scientists and evolutionary psychologists, including Gowaty (1982), Clutton-Brock and Parker (1995) and others, prefer to use the term 'forced copulation' for non-humans. Together with Thornhill and Palmer (2000), we do not believe such a distinction to be necessary. If qualification is required — such as human versus non-human rape — then we should apply it, just as we would in any other case.

6.4.1 Theories of Rape in Evolutionary Psychology

Evolutionary psychologists often view rape as a *reproductive strategy*, one that competes with consensual sex. According to parental investment theory, females who invest more in their offspring will choose to mate less frequently than males, which can lead to a conflict of interest between the sexes (Clutton-Brock and Parker, 1995). Males are faced with three evolutionary possibilities: they can make a greater parental investment, they can outcompete other males in securing more consensual matings, or they can secure more matings by committing rape. In our simulations here, we fix the parental investments and investigate the evolution of the other two strategies.

There is at least one significant pre-condition for the evolution of rape, namely that it be physically *possible* in the species. For example, there needs to be a pre-adaptation that makes males physically stronger than females (as Brownmiller and Mehrhol 1992 note is the case amongst humans), so that it is possible for the male to restrain the female during mating. As a corollary to this, females must not then evolve a defense against rape (Clutton-Brock and Parker, 1995). In particular, if females can evolve to become physically stronger than males or if they can evolve a means of eliminating unwanted sperm, then rape will be defeated as an evolutionary strategy (Gowaty and Buschhaus, 1998).

As we will see, these conditions hold by design in our simulations, but just how biologically plausible are they? A male physical superiority will hold in many sexually dimorphic species as a consequence of male-male competition and aggression, which is both plausible and commonly observed. There will be an absence of an evolved defense when its evolution would be more costly than simply permitting conception. Whether this is plausible warrants some discussion.

As we noted, defense mechanisms can take two main forms: females may either evolve greater physical strength or evolve to eliminate sperm. In the first case, the strengths of both sexes will coevolve in an arms race. However, males can always stay one step ahead in this race, simply because they can spend less on parental investment and more on strength (Clutton-Brock and Parker, 1995). Hence, this option will not normally affect the evolution of rape. On the other hand, sperm elimination would be both cheap and physically plausible — and the ducks in Gowaty and Buschhaus's (1998) study provide an example of such a defense mechanism. Of course, this may again launch a co-evolutionary race, with males developing means to overcome the defense. For simplicity, we do not account for these possibilities in our simulations, however the development of female defenses against sperm is certainly an important possibility that deserves further attention.

6.4.2 The Controversy over Evolutionary Accounts of Rape

Thornhill and Palmer (2000) present an account of rape from the perspective of evolutionary psychology. They describe the two main ideas on how the act may have evolved amongst humans: 1) the direct adaptation hypothesis and 2) the by-product hypothesis.[5] The direct adaptation hypoth-

---

[5]They also note the possibility of a drift hypothesis, but immediately dismiss it as too

CHAPTER 6. EXPERIMENTS IN ETHICS 183

esis holds that, in our prehistoric environment of evolutionary adaptation (EEA), rape was heritable and that individuals who performed rape were fitter than those who did not, at least under some real and recurrent circumstances. On the other hand, the by-product hypothesis holds that rape is a side-effect of *other* adaptive features of human behavior, but that rape itself has no effect on fitness. Both hypotheses make many of the same predictions, including the following:

1. Only men will evolve to rape.

2. The main victims of rape will be fertile women.

3. The main perpetrators will be men with low social status.

4. Perpetrators will target vulnerable individuals.

We only investigate the first of these predictions, though we certainly consider the others to be potential topics for future simulation work. Furthermore, the evolved rape we investigate is quite obviously adaptive: by design in our simulations rape is a heritable trait and the results we present make it clear that rape has an effect on fitness. Hence, of Thornhill and Palmer's two hypotheses, our simulations are most closely related to the direct adaptation hypothesis.

Both of Thornhill and Palmer's hypotheses have attracted many criticisms, the main one being that they are unfalsifiable (see, for instance, Coyne 2003). Certainly the hypotheses as put are vague and generic, but this does not mean they are unfalsifiable,[6] any more than similarly vague early claims of the adaptive value of antlers or peacocks' tails. Hypotheses that begin life as vague often mature to the specific and then end their lives as falsified. If new scientific hypotheses were routinely killed off for vagueness, there would be no next generation for science. To be sure, as we noted, both of the hypotheses yield many of the same predictions, and so distinguishing between them *non-experimentally*, as with social statistics, is bound to be difficult. We are in the more fortunate position of being able to conduct experiments, making the confirmability of the distinct hypotheses obvious.

Before we proceed to experimental confirmations and disconfirmations, we make a few remarks about the recent controversies over evolutionary

---

improbable.

[6]Or, far more appropriately, *unconfirmable*. Karl Popper's scientific methodology, based upon falsifiability, has long since proven itself to be degenerating relative to, for example, Bayesian methodology (see, e.g., Howson and Urbach, 2006).

psychological studies of rape. In a sense, the debate over rape is simply a more divisive version of the debate between evolutionary psychology and social science in general, which we discussed in §2.3.3. Evolutionary psychologists such as Thornhill and Palmer claim that the cause of some sex-based behavioral inequalities are heritable, while many social scientists disagree. They instead suggest that behavioral inequalities exist for two reasons: 1) differing (genetically determined) sexual anatomies and their non-evolutionary consequences, and 2) the cultural development of social and economic differences between the sexes (Eagly and Wood, 2003, p. 294; also, see Brownmiller and Mehrhol 1992).

Clearly, as evolved, social animals there must be some *combination* of cultural and genetic factors that explains our behavior, including differences in behavior that are sex linked. The idea that nature and nurture are *at odds* or that for one of them to have explanatory value the other must not is egregiously naive. We do not have enough information today to decide which combination of social and evolutionary models reflects reality — for rape or for other behavior. The existence of multiple competing models is common in science, but what complicates matters here is the intense political interest in rape. And there lies the nub of the problem: the social policies for rape that the parties propose do not generally concur, nor do their beliefs about how the world should be. Indeed, Thornhill and Palmer's (2000) most controversial claim is probably that we have implemented dangerous policies and given dangerous advice by ignoring evolutionary models of rape.

Thus the conflict is not primarily over standards of scientific rigor nor over the evidence required before a scientific proposition should be accepted as the truth, which is often how the debate is framed by both sides. Rather, the conflict is over what role a limited scientific understanding should play in the formation of public policy. Our simulations are far removed from models of specifically *human* behavior, let alone human public policy, so thankfully this conflict does not affect our simulations and its results. It is certainly an open possibility, however, to direct simulations of this type more directly at human behavior and human social functioning and so to engage them in these debates.

6.4.3 The Unethical Nature of Rape

No one questions the immorality of rape. Few acts are accorded this universal status; many people can imagine plausible situations that would justify theft or assault or killing, but rape does not seem to admit any such situation. Why is rape unique amongst these acts in being categorically unethi-

CHAPTER 6. EXPERIMENTS IN ETHICS 185

cal?

One approach to answering such questions is to consider variations of rape that are not obviously unethical. In biology there are a number of species whose mating act appears to be non-consensual, for example with elephant seals. Copulation among elephant seals is forced, and is characterized by the females struggling to flee. This is not obviously unethical, and it is not obviously rape, since there is no alternative means of copulation (Estep and Bruce, 1981). The preferences of female elephant seals are directly affected by males forcing copulation, by their being denied — after all the females are *trying* to flee. But clearly the species cannot evolve to the point where the females *succeed* in exerting their wills, not without some fairly dramatic changes elsewhere in their behavior.

We suggest that the central issue in the unethicality of rape is well captured by utilitarian considerations. An individual's consent is an expression of that individual's preferences. By violating consent, rape overrides an individual's preferences which, by definition, will yield a lower utility for that individual and, therefore, in the more plausible cases, collectively. In other words, the feminist account of rape as inherently an expression of power or dominance over the victim is correct, and that expression being a denial of the victim's ability to be guided freely by its own preferences is what renders the act unethical.

Whether there is the option of consensual sex is not to the point for the morality of non-consensual sex. It is relevant to the options available to evolution, and it may be relevant to whether the label of 'rape' is appropriate. In any case, in our simulations we have allowed for both consensual and non-consensual mating to evolve.

The Disutility of Rape

Rape can produce harm in several different ways. For the victim, harm can be either temporary or long term. In many cases, the long-term psychological harm that a victim suffers will be far more severe than the immediate physical harm. Thornhill and Palmer (2000) have put forth an adaptive explanation for this psychological harm, though it has met with some criticism (for example, see Kimmel, 2003, Coyne, 2003). While we do not consider the possibility in our simulations, harm may also extend to the victim's relatives and neighbors. It may even extend farther in society, creating the climate of fear that Gowaty and Buschhaus (1998) describe. There is also the risk to the perpetrator of being punished, or others being punished in error.

Even were all these forms of harm absent, we would still call the behavior rape. This may seem like a pointless concern — many may consider rape and the harm it causes inseparable. But separation is very much to the point of our investigations — psychological harm in response to rape need never have evolved at all. One question then is, in the absence of direct harms, whether rape would still be unethical given its other consequences. This is the question we explore in the last of the experiments involving rape in this section.

6.4.4 Simulation Design

Table 6.4 summarizes the important general simulation parameters that are distinctive to the rape experiments while Table 6.5 gives the health effects and utilities of both reproductive and non-reproductive actions. The main variable parameters include the probability of preventing a rape, parental investments after rape and mating, and the actions parameter itself.

Rape Prevention Probability (rpp). The rpp parameter specifies the probability that a rape attempt will be repelled. An agent suffers a substantial loss of health if it cannot prevent a rape; otherwise, it only loses a small amount of health. In our simulations rpp is either 0.9, 0.75 or 0.5. These probabilities produce a wide variety of sustainable rates for rape and consensual mating: at lower probabilities, consensual mating disappears; at higher probabilities, rape disappears.

Rape Action. Rape is a separate action that any agent can choose. If an agent chooses to rape, a random agent is targeted from its neighborhood for its victim. For the targeted agent, if sampling the rpp returns a failure to prevent the rape, the victim derives a large negative utility (-70 health units) and the rapist derives a small positive utility (5 health units). Furthermore, if one of the agents is female and the other male, and the female has enough health, she will conceive and give birth — regardless of whether she is the perpetrator or the victim. The female will then invest in the offspring on her own.

By contrast, if sampling the rpp results in the rape being prevented, the targeted agent experiences a much smaller negative utility (-10 health units) and the female involved will not conceive. In addition, the rapist suffers a large negative health effect (-60 health units) and utility (-15 health units), i.e., it is punished.

CHAPTER 6. EXPERIMENTS IN ETHICS

Parameter	Value(s)
Rape prevention probability (*rpp*)	0.9 0.75 0.5
Actions and their initial probabilities	**Eat** $\frac{1}{3}$ **Mate** $\frac{1}{3}$ Rest $\frac{1}{12}$ Walk $\frac{1}{12}$ Turn $\frac{1}{12}$ **Rape** $\frac{1}{12}$
Observations	Age, Health, Sex, Local food density, Local population density, **Mate requested**
Board size	40x40
Maximum entities per cell	1
Neighborhood size	7x7
Initial population	800
Health required for reproduction	200 health units
Parental investment	Equal to the health effect of conceptions, given in Table 6.5 below
Genome type	Production rules (7 fixed rules)
Food distribution function	Seasonal (period: 60 cycles, magnitude: 60 food units, mean: 130 food units)
Action probability mutation	Initial mutation rates: $N(0, N(0.01, 0.001)^2)$, Meta-mutation rate: $N(0, 0.001)$
Condition value mutation	Initial mutation rates: $N(0, N(0.01, 0.001)^2)$, Meta-mutation rate: $N(0, 0.001)$

Table 6.4: Parameters of the simulations containing rape.

	Success		Failure	
	Utility	Health	Utility	Health
WALK	5	-6	0	0
TURN	1	-2	–	–
REST	2	1	0	0
EAT	10	~140	-10	-10

MATE

Outcome	Utility	Health
Request accepted, birth	15	[-300,-240,-180,-120,-60]
Request accepted, no birth	15	-16
Request denied	0	0
Cannot find mate	-10	-10

RAPE – Victim

Outcome	Sex	Utility	Health
Rape, birth	F	-70	[-590,-470,-350,-230,-110]
	M	-70	-10
Rape, no birth	F	-70	-10
	M	-70	-10
Rape attempt prevented	F	-10	-10
	M	-10	-10

RAPE – Rapist

Outcome	Sex	Utility	Health
Rape, birth	F	5	[-590,-470,-350,-230,-110]
	M	5	-10
Rape, no birth	F	5	-10
	M	5	-10
Rape attempt prevented	F	-15	-60
	M	-15	-60

Table 6.5: The utilities and health effects for the rape simulations. (A dash indicates an impossible outcome.) On births the health effect equals parental investment. The parental investments for mate and rape are matched — for example, when parental investment after a mate is 300 health units, the female investment after a rape is 590 health units. Note that mating is identical for males and females, including subsequent investments.

CHAPTER 6. EXPERIMENTS IN ETHICS 189

As described above for our suicide experiments, in our rape experiments, we ran simulations in identical configurations both with and without rape enabled so that we could compare the ethical consequences of the two cases.

Consensual Mating Action. Consensual mating occurs in the same way as in other simulations.

Besides rape and consensual mating, the list of actions for these experiments includes eating, walking, turning and resting.

Parental Investment. For most of the experiments, males and females invested equally after *mating* — that is, investment was symmetrical. In the first experiment, there were five run sets for each *rpp*, each with a different level of parental investment after mating, ranging from 60 through 120, 180, 240 and 300 health units. Later, we relaxed the assumption of investment symmetry by increasing the female's investment to see the effect on sexual dimorphism and reproductive behavior.

Investments after rape were handled a little differently. Male parental investment remained fixed at 10 health units, which represented their minimum (gametic) costs. Female investment after rape was set to five different levels: 110, 230, 350, 470 and 590. These levels corresponded to the amount of investment needed in those runs to produce an offspring with the same health as an offspring born from consensual mating.

Observations. Agents are able to make the following observations: self-health, self-age, local-food-density, local-population-density, self-sex and mate-request. We introduced the self-sex observation because of its obvious importance to these experiments; we introduced mate-request to ease the evolution of consensual mating.

Health and Utility. The health effects and utilities associated with each action outcome are shown in Table 6.5. If a reproductive act leads to an offspring, then the parental investment in that offspring is equal to the magnitude of the health effects in this table. As always, it is only the qualitative relationship between these numbers that is of interest.

Production Rules. We used production rules, as described in §4.6.1, for these experiments, with rules tested against their observational conditions in random order.

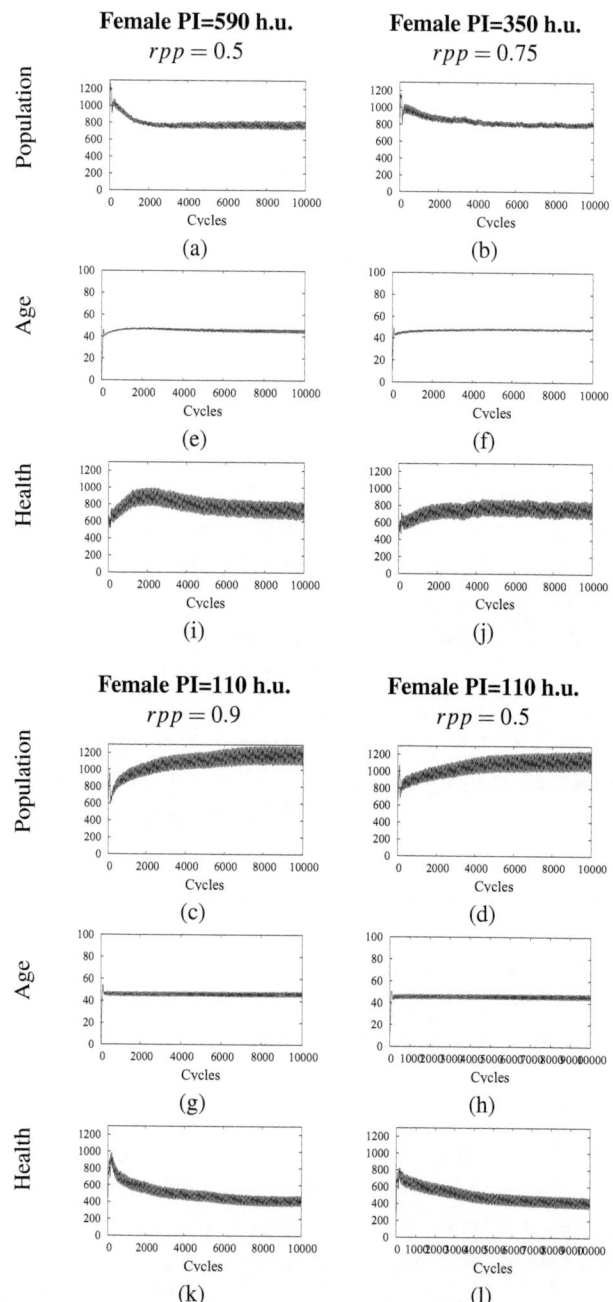

Figure 6.2: Average (a-d) populations, (e-h) age and (i-l) health for four different run sets. The labels across the top indicate the *rpp* and female investment used, per column.

CHAPTER 6. EXPERIMENTS IN ETHICS   191

6.4.5 Basic Demographics and Orientation

Let us begin by getting a feel for the simulations, as we did in the previous chapter. Figure 6.2 shows demographics for four of nine run sets from the first experiment; the graphs cover population numbers (Figures 6.2a-d), average ages (Figures 6.2e-h) and average health (Figures 6.2i-l). In the first two graphs (Figures 6.2a and 6.2 b), we can see that the population grew quickly at first, but then declined until settling at a lower size. In contrast, Figures 6.2c and d present populations that grew and then remained large but variable. The female investment for both of these graphs was the same (while *rpp* varied), therefore it looks as though the lower female investment was the main reason for the larger populations. Examining Figures 6.2i-l reveals lower average health in these cases. What is happening here is simply that agents need less food to sustain themselves and reproduce, so the food in the environment can sustain a greater number of agents.

By contrast, the average age in each of the simulations (in Figures 6.2e-6.2h) is mostly unremarkable, with one small exception. The lowest female investment produces the noisiest average age, which is as we would expect, given the greater population variability in those simulations.

6.4.6 Experiment: The Evolution of Rape and Dimorphic Behavior

Having looked at some of the effects of *rpp* and female investment on general demographics, we now examine the effect of these same parameters on the rate of attempted rape. Figure 6.3 shows the evolution of male and female rape rates.[7] Figures 6.3a-c are the results of three run sets with the *rpp* parameter set to 0.9, 0.75 and 0.5, respectively for each graph. In each case, female investment was at its highest level of 590 health units. From these graphs, we can immediately see the strong effect that *rpp* can have on rape evolution.

Figure 6.4 gives a view of what happened in all the run sets. Again, Figures 6.4a-c represent simulations in which the *rpp* parameter is set to 0.9, 0.75 and 0.5, respectively. Now, however, within each of these graphs, the horizontal axis further subdivides the simulations into five groups, corresponding to distinct female investment levels. Thus, the pair of bars labelled '350' in Figure 6.4b represents the male and female rape rates in simulations where the $rpp = 0.75$, and the female and male investments

---
[7] All the results are averages of 30 simulations.

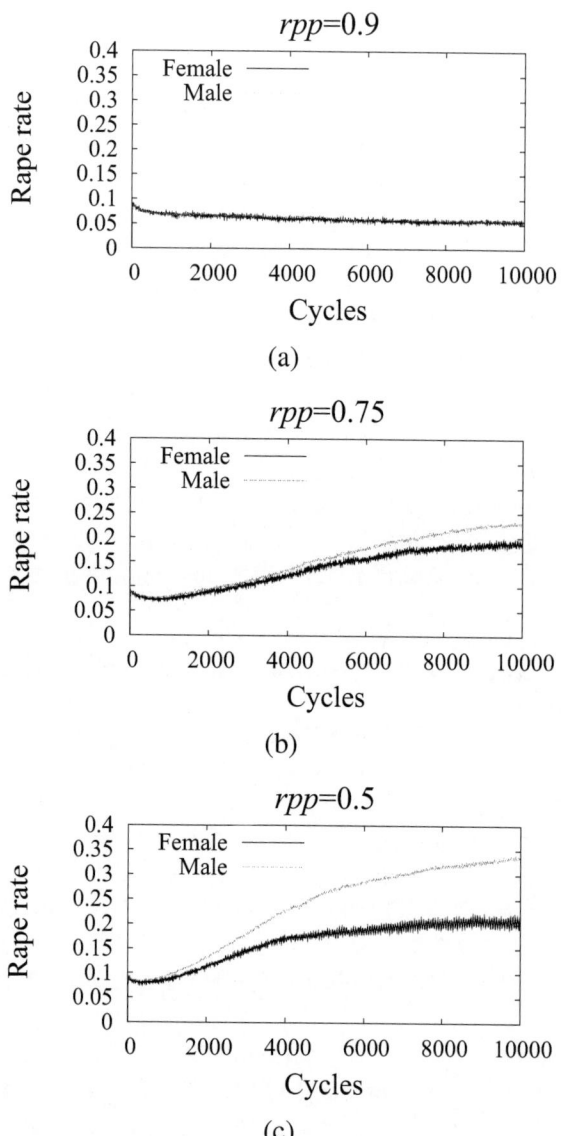

Figure 6.3: Female and male rape rates evolving over time when female parental investment is 590 health units, for rpp = (a) 0.9, (b) 0.75 and (c) 0.5.

CHAPTER 6. EXPERIMENTS IN ETHICS 193

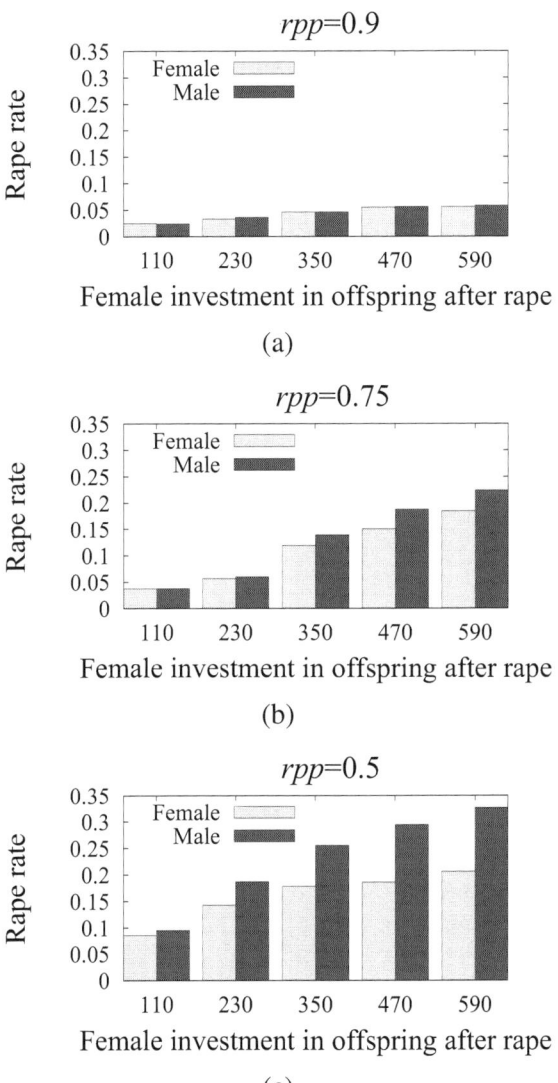

Figure 6.4: Female versus male rape rates for different levels of female parental investment and a fixed male parental investment of 10 health units, for *rpp* = (a) 0.9 (b) 0.75 and (c) 0.5.

are 350 and 10 health units respectively. Each bar in the graphs represents the average rape rate for the last 2000 cycles.

To round off the results, Table 6.6 summarizes the rates of eating, mating and rape for the various levels of *rpp* and female investments. Most of the differences between males and females shown in this table are statistically significant; we have italicized those that are not. We have omitted resting, turning and walking from this table for clarity and because they showed no notable differences between the sexes.

rpp=0.9

FPI	Eat			Mate				Rape		
	F	M	Diff	F	M	Diff	Accept	F	M	Diff
110	0.547	0.542	-1.0%	0.266	0.271	1.8%	0.440	0.025	0.025	-1.1%
230	0.487	0.484	-0.7%	*0.311*	*0.310*	*-0.3%*	0.361	0.033	0.034	1.8%
350	0.422	0.414	-1.9%	0.362	0.365	0.7%	0.206	*0.046*	*0.046*	*-0.6%*
470	*0.377*	*0.376*	*-0.2%*	*0.384*	*0.383*	*-0.2%*	0.191	0.055	0.056	1.0%
590	0.361	0.363	0.5%	0.392	0.395	0.7%	0.158	0.056	0.058	3.2%

rpp=0.75

FPI	F	M	Diff	F	M	Diff	Accept	F	M	Diff
110	0.562	0.550	-2.2%	*0.245*	*0.247*	*0.7%*	0.495	*0.038*	*0.038*	*-0.3%*
230	0.518	0.509	-1.8%	0.252	0.264	4.7%	0.289	0.057	0.060	5.9%
350	0.436	0.402	-7.7%	0.266	0.269	1.1%	0.179	0.119	0.139	17.1%
470	0.421	0.368	-12.6%	0.239	0.260	8.6%	0.149	0.151	0.188	24.2%
590	0.415	0.333	-19.7%	0.210	0.242	15.0%	0.130	0.185	0.224	21.0%

rpp=0.5

FPI	F	M	Diff	F	M	Diff	Accept	F	M	Diff
110	0.590	0.573	-2.9%	*0.180*	*0.181*	*0.6%*	0.381	0.086	0.095	9.9%
230	0.597	0.536	-10.3%	0.108	0.116	7.5%	0.387	0.143	0.187	30.9%
350	0.596	0.487	-18.3%	0.071	0.085	19.1%	0.159	0.178	0.255	43.4%
470	0.568	0.447	-21.3%	0.085	0.089	4.8%	0.130	0.186	0.294	57.9%
590	0.549	0.394	-28.2%	0.074	0.100	35.4%	0.111	0.206	0.327	58.8%

Table 6.6: A table of the eating, mate and rape attempt rates over the last 2000 cycles for simulations with different (fixed) female investment and *rpp* (0.9, 0.75 and 0.5). FPI reports the female parental investments. The F columns give frequencies for females and M those for males. Diff reports the male minus the female frequencies as a percentage of the female frequencies. Accept shows the rate of accepting mate requests. Italicized entries failed statistical significance tests at the 0.001 level.

6.4.7 The Evolutionary Stability of Rape

We can immediately see from the graphs of Figure 6.3 and 6.4 that rape can be evolutionarily stable. Examining the graphs in Figure 6.3 in particular,

CHAPTER 6. EXPERIMENTS IN ETHICS 195

we can see that rape in the $rpp = 0.5$ and $rpp = 0.75$ run sets eventually evolves to a rate that is much higher than its initial rate. From this, we can infer that most, if not all, of those simulations produced evolutionarily stable rape.

It is easy to see how rape in the simulation can be evolutionarily stable. Rape and mating can be viewed as evolutionarily competing strategies for reproduction. The more effective rape is in producing viable offspring, the less fitness benefit (or the greater fitness cost) an agent will derive from consensual mating, and vice versa. Like mating, rape can directly increase the fitness of the perpetrating agent by producing viable offspring for the perpetrator. Furthermore, it can do this without the complication of obtaining consent. However, this does not guarantee that it increases fitness, since rape also comes with the potential for failure and severe punishment.

Comparing the graphs in Figure 6.4, we can see that the fitness effect of rape is boosted by a lower *rpp*, as we would expect given the greater chance of producing a child. This is true for both males and females, though more so for males. It is interesting that this also holds for females, since rape is so much more costly for them — even when they are the perpetrators; for example, compare the male and female rape costs in Table 6.5. From Table 6.6, we can see that when the $rpp = 0.5$ and female investment after rape is 590 health units (compared to 300 health units after mating), that females strongly prefer rape to mating, with a rape rate of 0.206 compared to a mating rate of 0.074.

Thus, despite the fact that females are investing twice as much after rape for an offspring that is of the same health, they still prefer rape to mating. It is likely that the higher health cost of rape is being offset by a higher probability of producing a child via rape. Note that the probability of conceiving after requesting a mate will depend on whether the other agent rejects the request to mate; on the other hand, the probability of conceiving after rape will depend on the *rpp*.[8] From the *Accept* column of the final row in Table 6.6, we can see that the rate at which agents reject mates is 0.889, which is well above the *rpp* of 0.5.

6.4.8 The Emergence of Sexual Dimorphism

One of the most striking features of these results is the sexually dimorphic behavior that evolves. As each graph shows, dimorphic rape behavior corre-

---

[8]In both cases, conception will also depend on the agents both having the required health.

lates with the level of female investment after rape. As female investments increase, the sexual difference in the cost of rape also increases, since male investments after rape remain constant at ten health units. This indirectly leads to sexual differences in rape rates because, as we will soon see, females change their strategy to eat more and rape less.

The level of female investment on its own is not enough to produce noticeable sexually dimorphic behavior. When the *rpp* is high (i.e., when the $rpp = 0.9$ in Figure 6.4a), sexual dimorphism is not clear, even at the highest female investment level. With a decrease in the *rpp* to 0.75, dimorphism becomes obvious at levels of female investment of 350 health units and greater; while $rpp = 0.5$ delivers obvious dimorphism at all the female investment levels.

The influence of the *rpp* on sexual dimorphism is perhaps surprising. The correlation indicates that decreasing *rpp* affects the fitness of rape differently for each sex (although the effect is always positive). The sex difference appears to be due to an interaction between the health of each sex and the *rpp*. In particular, as the *rpp* decreases, the average fitness benefit of rape for any perpetrator increases. This increases the incidence of rape, which, due to the different investment levels by each sex after rape, exacerbates the differences in health for each sex. This, in turn, leads to rape being less effective for females, exaggerating the evolution of dimorphism in rape behavior. In short:

1. lower *rpp* leads to a higher rape rate;

2. which leads to females becoming much less healthy than males;

3. which leads to rape being a less adaptive option for females than males.

Given the existence of dimorphic rape behavior, dimorphism in other behavior must also manifest itself; if agents perform less of one action, they will have to perform more of another (even if that other action is simply resting). Looking at Table 6.6, we can see that there is, in fact, notable sexual dimorphism in eating and mating behavior. We can easily understand the dimorphism in eating behavior given the different health levels of each sex. Females suffer greater costs from rape, leading to a higher cost of living overall, and need to compensate by eating more, as is evident when $rpp \leq 0.75$.

Dimorphism in mating behavior also depends on sex differences in health. While the act of rape is itself dimorphic, costing much more in

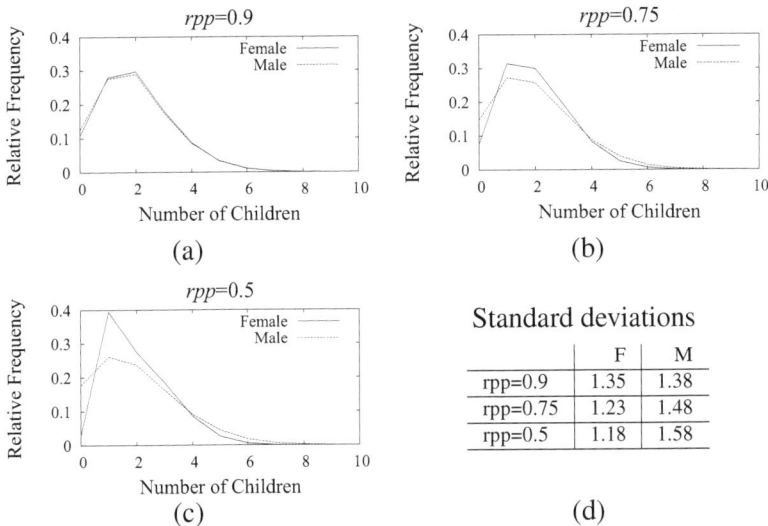

Figure 6.5: Frequencies of the number of children per agent (grouped by sex) in simulations with *rpp* equal to (a) 0.9, (b) 0.75 and (c) 0.5. (d) and the standard deviation for each frequency polygon (averaged over 10 simulations).

health for females than males, mating costs remain the same for both sexes. However, once rape becomes a part of a mixed reproductive strategy, the average health of each sex diverges, and mating will acquire a distinct adaptive value for each sex. This is an example of an existing evolved dimorphism causing further divergence in the behavior of each sex. In consequence, smaller *rpp* and larger female investment levels produce substantial dimorphisms in mating.

Dimorphic Reproductive Variability

Parental investment theory predicts that if there are differences between the sexes in their investments, the sex that invests more (ordinarily females) will be the limiting factor in sexual reproduction (see §5.4), i.e., females will have easier access to mates than males. In a stable population, females *and* males will, of course, have 2 children on average. However, the *variation* in the number of offspring for females will generally be lower than that for males.

This prediction holds true of our simulations. Figure 6.5 shows the dis-

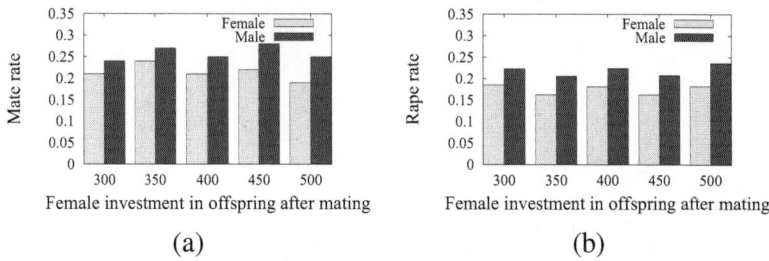

Figure 6.6: Female/Male (a) mate and (b) rape rates for different female and male parental investments after mating ($rpp = 0.75$). The male investments are equal to 600 minus the female investment.

tribution of the number of children born to parents by sex. The graphs and table are taken from 10 simulations with female investment levels of 590 health units. We can see that there is not much difference between the sexes at $rpp = 0.9$, when the frequency of rape is low. However, as the rpp decreases in Figures 6.5b and 6.5c, reproductive variability diminishes for females and increases for males (Figure 6.5d). At $rpp = 0.5$ females are almost certain to have one child, while males run a sizable risk of not having any children (with 0.18 probability).

Introducing Distinct Investments after Mating

The preceding simulations assume that males and females invest the same amount after consensual mating; now we relax this assumption. Figure 6.6 shows the rates of mating and rape from simulations with $rpp = 0.75$ and female investment after rape is fixed at 590 health units. Each graph shows the behavior rates at the given levels of female investment after mating (marked on the horizontal axes). Male investment after mating in these graphs is 600 health units minus the female investment after mating, which is, like female investment, constant for any given simulation. The bar graphs are again averaged over the last 2000 cycles of the simulations.

Figure 6.6b tells us that asymmetrical investments after mating do not significantly affect the rates of rape, for either sex. It is plausible that rape rates would be affected, since there is now less that separates mating from rape from the perspective of both sexes. However, the rpp must still have a larger influence on the adaptive value of each of the two reproductive actions than health effects after mating — which, even at their most exaggerated, are less severe than the health effect differences from investment

CHAPTER 6. EXPERIMENTS IN ETHICS   199

after rape. Unfortunately, simulations with greater levels of female investment after mating turn out not to be stable, as females can no longer sustain the large hit to their health. Nevertheless, it is clear from the simulations here that altering after-mating investments for females does not push the dimorphism of rape in any particular direction.

Figure 6.6a shows the mating rates for the same simulations. Dimorphic mating behavior increases with larger female investment, with females mating less than males. Since dimorphic rape rates are unaffected by this change, females are opting to request fewer matings overall. Instead of choosing to mate or rape, females will eat, so as to compensate for the bigger hit to their health — for instance, there is an increase of around 0.025 in the probability that a female will eat in the simulations where females invest 500 health units.

Reducing Sexual Dimorphism

In order to test the evolutionary boundaries of the dimorphism in reproductive behavior and its robustness, we tried various alternatives to reduce the sexual dimorphism without reducing the difference in parental investment. Figure 6.7 shows the results for two such alternatives. The graphs in Figures 6.7a and 6.7c are taken from simulations in which we *compensated* females for their investment after rape, by providing them with a high utility from having an offspring, but only if they initiated the rape. Figures 6.7b and 6.7 d shows the effect of decreasing the *efficiency* of male eating — in particular, when males absorbed just 40% of the health available in food. Our aim with these simulations is to neutralize the health differences between the sexes caused by rape.

Figure 6.7a shows that even this fairly implausible scenario of utility compensation is incapable of reducing sexual dimorphism in rape behavior: it has had no effect on the dimorphism of the rape rates. It may seem curious that, even though rape costs almost the same for both sexes to perpetrate, males perpetrate rape much more frequently. But as victims females endure a much greater burden than males, so their average health is much less, as shown in Figure 6.7c. This, coupled with the fact that rape attempts that are repelled result in a heavy punishment for the perpetrator, means that rape remains a much less adaptive option for females than males.

The male inefficiency simulations of Figure 6.7b show an actual reduction of differences between the sexes. Looking at Figure 6.7d, we can see that the reduction is due to a shift in health differences between female and male agents, with females now being healthier. Of course, rape is still a

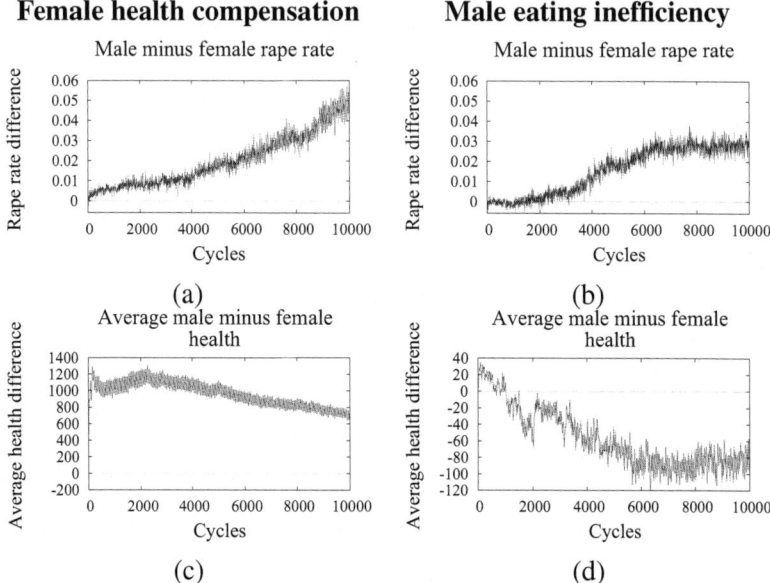

Figure 6.7: Rape rate differences between the sexes at $rpp = 0.75$, when (a) females gain sufficiently high utility for rape attempts to compensate for high investment and (b) males are limited to absorbing 40% of the health in food. Also, (c) and (d) the male minus female difference in average health for (a) and (b), respectively.

much more costly behavior for females, so the dimorphism still does not disappear.

6.4.9 The Genetic Causes of Rape and Sexually Dimorphic Behavior

Over the previous sections, we have looked at situations that produce evolutionarily stable rape behavior and, in particular, situations that produce or affect sexual dimorphism in rape, mating and agent behavior in general. Now we will look at the genetic causes of sexually dimorphic rape behavior within individuals. We can do this by examining the sex-linked differences in the probability of selecting the rape action, i.e., by comparing action distributions conditioned upon self-sex. Figure 6.8a shows that difference for the $rpp = 0.75$ simulations, with female investment at 590 health units, from §6.4.6. Clearly, sex-linked rules do indeed contribute to the dimorphism.

CHAPTER 6. EXPERIMENTS IN ETHICS

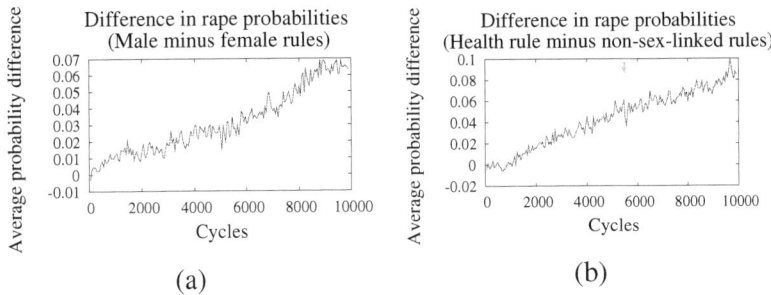

Figure 6.8: (a) The average difference between male and female probability of rape in the genome for the *rpp*=0.75 runs from Figure 6.4(b). (b) The average probability of rape in high health rules minus the average probability of all other non-sex-linked rules.

We can also isolate the effect of the self-health rule and check its contribution to the rape-rate difference. Since this rule contains a greater than operator and, on average, the health threshold evolved to around 600 health units, it is a good indicator of behavior given higher levels of health. To determine this rule's effect on rape rates, we take the difference between this average and the average for the remaining rules, ignoring the self-sex rules:

$$P_\delta = \overline{P}_h(r) - \frac{1}{n} \sum_{i \neq h,m,f} \overline{P}_i(r) \qquad (6.1)$$

where $\overline{P}_i(r)$ is the mean probability of rape (r) in the ith rule; h, m and f are the indexes of the self-health, male and female rules, respectively; and n is the number of rules. Figure 6.8b shows this statistic for the experiments from §6.4.6. We can see a more significant difference evolving here than in Figure 6.8a, which would indicate that the health rule makes a slightly larger contribution to the difference in rape rates. This confirms what we have seen in earlier sections, in which the *rpp* affects the level of dimorphism due to its effect on the average health.

Sex-only Genome

We also looked for sexually dimorphic behavior when agents are restricted to conditioning on self-sex alone, i.e., there is only one behavioral rule being used per agent, the one which matches its gender. From our prior results we expected that dimorphism would still evolve, though perhaps to a lesser extent.

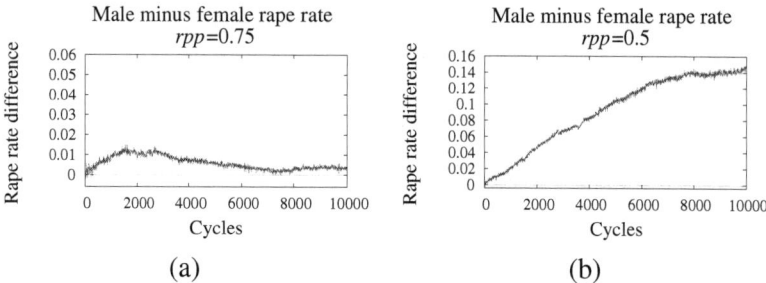

Figure 6.9: The rape rate difference between sexes for (a) $rpp = 0.75$ and (b) $rpp = 0.5$ when sex is the only observation allowed.

The graphs in Figure 6.9 are the result, showing, at most, negligible dimorphism in the $rpp = 0.75$ simulations, but for $rpp = 0.5$ the difference is *greater* than that in the earlier simulations. The explanation of these outcomes is unclear. Certainly, the lack of other rules in the genome, and the concomitant inability to determine one's circumstances, makes a common strategy desirable in the former case. However, when $rpp = 0.5$, since females are generally in low health and cannot change behaviors dependent upon changes in health, they must maintain a strategy primarily of eating, while males are substantially healthier and so freer to pursue alternatives.

6.4.10 Experiment with the Ethical Consequences of Rape

The question of the actual ethicality of rape amongst humans is so clear as to be hard even to raise. Victims, their families, friends and communities all are subjected to large negative utilities. Whatever happens to perpetrators, the ethical implications of their outcomes are dwarfed by the other consequences of their actions. The utilitarian verdict is plain. However, it is possible that rape is unethical in an even more fundamental sense. For even were these large negative utilities — the direct costs of rape — not to exist, it may be that rape would still be unethical from a utilitarian perspective.

In our first set of experiments here, represented in Figure 6.10a, we checked that rape is unethical as it stands in the simulations above, and we also examined the effect of the *rpp*. The lines in the graph represent the cumulative utilities for a set of simulations with the given parameters. The uppermost line plot represents simulations in which rape is not possible. This line clearly shows that disallowing rape altogether produces the most ethical environment for the agents, as we would expect. The plot line immediately below came from simulations in which the $rpp = 1$. That is, in

CHAPTER 6. EXPERIMENTS IN ETHICS

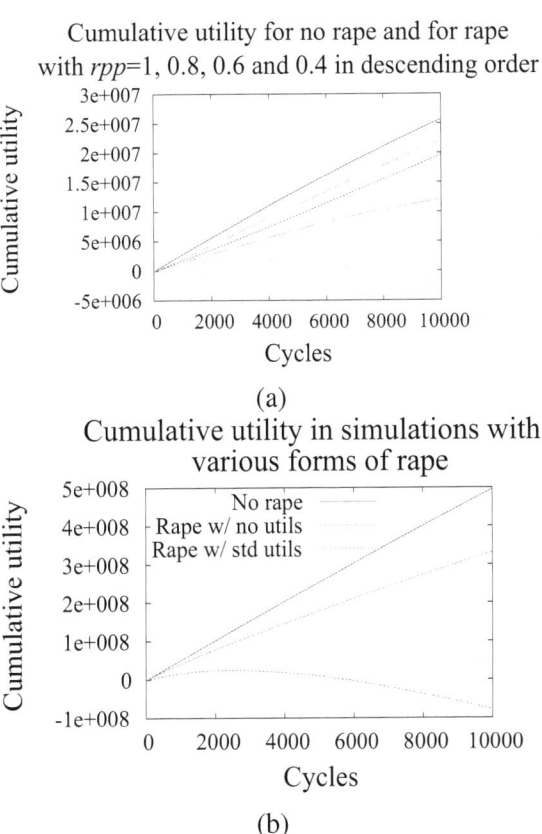

Figure 6.10: (a) Cumulative utilities for simulations with (top to bottom) (i) no rape, (ii) rape with an $rpp = 1$, (iii) rape with an $rpp = 0.8$, (iv) rape with an $rpp = 0.6$ and (v) rape with an $rpp = 0.4$; (b) cumulative utilities in simulations with (i) no rape, (ii) rape with all rape utilities equal to nil and (iii) rape with utilities per normal.

these simulations rape is an available action, but it is never successful and always punished. This, of course, reduces total utility by punishing those unhappy enough to attempt rape. As rape can hardly be fit, these will be actions induced by a mutation yet to be selected out of the genome. The remaining plot lines show the results of $rpp = 0.8$, $rpp = 0.6$ and $rpp = 0.4$, with a positive correlation between rpp and cumulative utilities. In general, these results show a growing utilitarian damage to the population as rape becomes a fitter option for reproduction: the more rape is rewarded in fitness terms, the more the strategy is adopted and the worse the utilitarian outcomes for all concerned.

It is interesting to see that the $rpp = 0.4$ simulations produced a plot of cumulative utilities that is almost parabolic for the duration investigated, which means that the total population utility continued to drop at every epoch. In fact, this was because the rape rate continued to increase smoothly throughout, supporting the point that the rpp and cumulative utilities are correlated through the mediation of the rape rates. The $rpp = 0.4$ simulations end up supporting perpetually negative utilities for the population as a whole, suggesting that no population at all would be preferable to the situation characterized by the simulation.

The results could have been different. After all, rape can result in the birth of offspring. If a child can expect to obtain positive utilities throughout life, this might compensate the negative utility the victim experiences from rape. However, there are two factors working against this possibility here: a high incidence of rape will lower the child's expected utility (along with everyone else's), and the population size is fairly steady regardless of the frequency of rape.

So, the rape in our simulation is unethical, and the higher its frequency, the greater its harm. However, it is unclear whether rape's unethical value is entirely due to the large negative utilities associated with being a rape victim. In our simulation these are given parameters of the situation, and not under evolutionary control. We decided to determine whether rape would remain unethical without these utilities.

Figure 6.10b shows the results of three sets of simulations. The top and bottom plot lines are from simulations that contain no rape and rape with standard utility settings, respectively.[9] The middle plot line describes a set of simulations in which rape produces no utilities at all for victims or perpetrators. Both simulations containing rape used $rpp = 0.4$. Utility-less rape

---

[9]These plot lines differ very slightly from those of the neighboring graph due to minor changes between the corresponding experiments.

produces a much more ethical world for the agents: we can no longer say their lives are not worth living. However, utility-less rape simulations continue to lead to worse outcomes than simulations without rape. This is down to the large health cost of rape for females, who, being less healthy, must adopt less optimal strategies than if there were no rape. Since the differential in health outcomes for rape is plausibly a feature common to the general circumstances of rape, this supports the idea that rape cannot be turned into even an ethically neutral event by neutralizing its direct utilitarian costs. It is unethical at a more fundamental level.

6.4.11 Summary

The first thing we note in summary is the critical role played by the *rpp*. When the probability of preventing rape is high, both the level of rape and the level of its sexual dimorphism that evolve are low. For humans, Thornhill and Palmer (2000) have estimated that the conception rate resulting from rape is about 0.02. This would be too low for rape to exhibit dimorphism under the conditions of our simulations. An alternative would be dimorphic rape behavior caused by indirectly related traits, such as dimorphic levels of strength or aggression. The need to invoke alternative causes of dimorphism would seem to support the by-product hypothesis.

We also saw that the difference in female and male parental investments after rape can have a substantial effect on the behavioral dimorphisms that evolve. An important possibility for future investigation is the evolution of counter-strategies by females, such as sperm elimination and increased strength, which may serve to reduce behavioral dimorphism.

The primary cause of sexually dimorphic behavior in our rape simulation appears to be health — that is, since females invest more in the resulting offspring, they are less healthy on average than males, and therefore prefer eating to copulating more frequently than males. An excellent proxy for health in many of the simulations is sex — and thus sexually dimorphic behavior is caused not just by the conditions an agent finds itself in, but also directly by an agent's sex. That is, sex-based inequality in behavior evolves.

The ethical investigations of rape show what we already know: rape is unethical. However, they do produce one novel result: even after eliminating the negative utility — the direct harm — associated with rape, it remains an unethical act. These simulations also underline the further critical point that what evolves is not always ethical; indeed, the fitter rape becomes, the worse off the world's inhabitants are.

6.5 Abortion

Many of the most heated ethical debates revolve around abortion, often centering on the moral status of the fetus. While on utilitarian grounds the future of the fetus is indeed a central issue, an exclusive focus on it ignores the consequences of abortion on everyone else, which can readily turn out to be more significant. This is also the perspective we take for the ethical experiments in this chapter.

Abortion does not just terminate a fetus; it terminates all further parental investment in it. The relation between parental investment and abortion is at the center of the evolution of abortion. Indeed, for our purposes here, we simply define abortion as an act that ends all parental investment for offspring conceived but not yet born.

Parental investment theory in evolutionary psychology provides a natural explanation for the evolution of abortion. For example, a pregnant mother that suddenly finds herself in difficult circumstances may benefit her fitness by saving her resources for future offspring who have a better chance of surviving. If this is so, it is unlikely that fitness and ethicality will diverge for abortion, as it did with rape. Indeed, these are the very evolutionary hypotheses we explore now.

6.5.1 Abortion and Evolution

Abortion takes two forms: spontaneous and induced. **Induced abortion** is a deliberate action, whereas spontaneous abortion is not. In our experiments we simulate one kind of abortion, sharing features of both. On the one hand, an agent can abort based on the current state of its health, which is most similar to **spontaneous abortion**. On the other hand, an agent can base abortion on changes in the environment that may take time to affect the agent, which more resembles induced abortion.

Below, we present several examples which suggest that abortion in the real world is under evolutionary control. The circumstances are often similar to those we will see in our simulations.

Examples of Abortion in Nature

Humans: Spontaneous Abortion. A healthy mother is susceptible to spontaneous abortion in at least 15% of cases (McBride, 1991), but there have long been suggestions that it may occur even more frequently (Roberts and Lowe, 1975). At least half of these cases are a result of chromosomal abnormalities, while other factors may include a family history of spontaneous abortion, the age of menarche, whether the potential father is a close relation, abdominal trauma and infection (Al-Ansary et al., 1995).

In at least the case of closely related parents, Verrell and McCabe (1990) have suggested that spontaneous abortion is, in fact, adaptive. In these cases, the offspring has an increased chance of carrying two harmful recessive alleles at the same locus, particularly in the genes responsible for the immune system. As such, children have an increased chance of expressing harmful recessives that are normally suppressed. The decreased chance of the child surviving may make abortion a good evolutionary strategy for saving parental investment for others and, so, increasing total fitness.

Humans: Induced Abortion. Induced abortion has only become safe and effective in the last century, although it was practised in earlier times (Devereux, 1954, Hrdy, 1999, p. 470). It therefore is implausible that it could have evolved directly. Even so, we may find abortion has many of the features of a directly evolved act, namely **filial infanticide**. Filial infanticide has clearly been part of our evolutionary history, as well as that of other animals (Hrdy, 1999). For pregnant females the effect of filial infanticide on parental investment is very similar to abortion.

Lycett and Dunbar's (1999) study lends support to the idea that induced abortion is performed in ways that would have been reproductively beneficial in our evolutionary past, implying that it can be a rational choice now for enhancing fitness. Using data from England and Wales for 1991, they found evidence for two main predictions derived from parental investment theory:

- Amongst younger women, those who were single were more likely to abort than those who were married.

- A preference for abortion declines amongst single women as they age, but increases for married women.

The theoretical argument for the first prediction is that unmarried women normally have less social and material support for raising children; by aborting, they conserve their resources for a future occasion when they may be combined with new sources of parental investment. The argument for the second is that, as the chances of single women finding a suitable partner declines with age, they will prefer single parenthood to no offspring at all; contrariwise, older married women aborting are most likely preserving their future investments for existing offspring.

Mice. In cases observed in the lab, a female mouse will abort her litter when introduced to a strange male (Bruce, 1960). Soon after, the female becomes sexually receptive again and mates with the strange male. Hrdy (1979) has suggested this is a female adaptation to the behavior of males, since an unrelated male often kills the female's offspring once born. The example clearly demonstrates a situation in which conditions change after conception; females cannot know beforehand that they will meet strange males.

Yucca Plants. Another example of spontaneous abortion can be found in yucca plants (Pellmyr and Huth, 1994, Huth and Pellmyr, 2000). The yucca plant depends on the yucca moth for its pollination. A female moth deposits her eggs in a yucca flower's ovary, and then places pollen that she has collected on the stigma of the flower, so that the flower will produce a sufficient number of seeds to feed her offspring. However, as Pellmyr and Huth (1994) note, the yucca plant occasionally aborts flowers when the flower is oversupplied with eggs in comparison to the number of seeds. Thus, when the survival and reproductive prospects of the flower are worse than what continued health investment would justify, the flower is aborted.

6.5.2 The Ethics of Abortion

So, there are clear fitness grounds for the evolution of abortion as a means of conserving the potential for parental investment in a fairly wide variety of circumstances. The ethical standing of abortion is, of course, quite another matter.

The Moral Status of the Fetus

The most politically contentious issue surrounding abortion is the moral status of the human fetus, or, indeed, embryo. For utilitarians (excepting perhaps Peter Singer; see §2.5), and our simulations, this issue depends upon the potential for the fetus to experience positive utility in the future, and it needs to be balanced against the expected utilities of everyone else.

By contrast to our position, the two most prominent views both tend towards a moral absolutism. The religious **pro-life** movement is explicitly based upon absolutism, with an unalterable sanctity asserted for human life, beginning at conception. In contrast, for many secular **pro-choice** supporters the main issue is when a human can be counted as a moral being, with the rights and sentient capacities of other people. For example, according to

Hrdy (1999, p. 392), this occurs when a human develops the "unique empathetic component that is the foundation of all morality," which can only happen in a social context. Therefore, life begins once an individual begins socializing, which Hrdy asserts is either at birth or a little earlier. Sumner (1981) places similar importance on socialization. However, Sumner suggests that the appearance of sentience is the "threshold" event, that distinguishes moral beings from beings that are simply alive. In any case, the interests of those who are granted the status of moral agents simply *trump* the potential interests of those who are not, which certainly does smack of moral absolutism. In utilitarianism the interests of everyone are meant to be accommodated as best as can be managed; no one is simply trumped.

In interpreting our simulations all individuals are taken into account. However, we simulate unborn agents without actions and with utilities. They have an interpretive voice in that their *potential* utility, as born and active agents, counts when we compare simulations in which they were aborted with matched simulations where they were not.

Utilitarian Perspectives on Abortion

From the perspective of utilitarianism, then, not all ethical issues of abortion are simply solved once we decide whether the fetus is a moral agent. The intricate pattern of the consequences of abortion is what matters. That is one task in which ALife simulation, with its capacity for exploring consequences across a wide range of ethical circumstances, can assist utilitarian thinking.

There are several existing utilitarian perspectives on abortion. For a prominent example, Peter Singer suggests that only those entities capable of feeling pleasure and pain ought to be counted in the utilitarian sum and that babies less than a month old, lacking consciousness, do not count (Singer, 1993). He argues that not only is abortion often acceptable, but that so too is infanticide in the first month after birth — although, only if the child has severe disabilities and can expect negative utilities throughout its life. Singer's views on consciousness have no serious basis in the neuroscience of consciousness (see, e.g., Janus, 1997 on prenatal experience). In any case, his conclusions are not likely to find any wide support in our society, even though there certainly are precedents for the cultural acceptance of infanticide (Hrdy, 1999).

Sumner (1981) suggests that we divide abortions between early and late term, based on when the fetus achieves sentience. Early term abortions would be entirely unregulated, while late term abortions would be permitted for women whose lives are at risk or when the child may have a severe disability. Much like Singer, Sumner implies that *infanticide* may also be allowed in the case of severe disability, so long as the infant has made no social attachments.

Despite their contrary claims, neither Singer's nor Sumner's positions are in fact consistent with utilitarianism. The division into early-term and late-term abortions based on a sentience threshold is *not* just a means of clarifying when the fetus begins to experience negative and positive utilities. It is primarily a way of writing off the *future utilities* of fetuses. In Sumner's (1981, p. 277) words, "Before the threshold there is (in the deontically relevant sense) no creature to harm and no right to life to be violated." Here Sumner makes explicit reference to the deontics to which utilitarianism is opposed. He decries the revisions of classical utilitarianism of *others*, such as average utilitarianism and "discount theories," in which entities that do not yet exist are given less weight in the utilitarian sum,[10] and yet Sumner himself appeals to non-utilitarian deontic principles. Perhaps these debates illustrate how easy it is to be seduced by the language of the debate, "the right to life" and "the freedom to choose", the language of the absolute rights of one to trump those of another. Utilitarianism proper, and its simulation here, stand in opposition to such ethical despotism.

6.5.3 Simulation Design

As we have seen then, abortion has evolved many times in nature, in plants and animals, including humans. While human abortions can be divided into two distinct types — spontaneous and induced — there is reason to believe that both are shaped in similar ways by evolution and that simulations like ours can shed some light on their evolutionary nature. While simulation cannot assist with an answer to the most contentious question about abortion's ethical nature — the moral status of the fetus — simulation *can* help with our understanding of the most important question about abortion's ethical nature — its consequences.

With this in mind, we can turn to the design details for the coming simulations. As before, we will discuss here only those aspects specific to the abortion experiments. For a summary of the parameters, the reader can refer to Tables 6.7 and 6.8. The latter table covers the health effects and utilities of all the actions, while the former table covers the remaining

---

[10]Such a discount, by the way, is implicitly required by classical utilitarianism, since *expectations* for the future are inevitably weaker the more distant that future is.

CHAPTER 6. EXPERIMENTS IN ETHICS 211

parameters.

For the main experiments (§§6.5.5 and 6.5.6) we varied three parameters: gestational investment (*gi*), after-birth investment (*abi*), and the food distribution function (*fdf*). While we have explored varying other parameters — including male investment, the length of the gestational period and the health effect of abortion — we found that variations in the three above to be the most relevant and interesting.

Variable Parameters

Parental Investment. There are two main forms of parental investment in these simulations: gestational investment (*gi*) and after-birth investment (*abi*). (There is also a third form, gametic investment, which is both small and fixed.) In these experiments, for simplicity only females contribute to *gi* and *abi*. We varied *gi* and *abi* through the values 0, 20, 150 and 300 health units in both the evolutionary and ethical experiments. Table 6.8 shows the parental investments made after successful mating (as health losses to the parent). Note that females and males invest 15 health units and 10 health units, respectively, in their gametes, in accord with the biological definition of the sexes.

Actions. In order to simplify our genetic analysis, we allowed at most three actions in these experiments: eating, mating and abortion.

Abortion. Agents can try to abort at any time; if an agent is not pregnant, then abortion merely results in self-harm. Whether pregnant or not, an agent loses 40 health units when trying to abort. When an agent is pregnant, abortion terminates the pregnancy and any gestational investment made to that point will be lost. Agents will, of course, save any future investment that would have gone into the offspring had they not aborted. We did not assign any utilities to the different outcomes of abortion, preferring to keep such matters neutral.

Food Distribution Function. The food distribution function (*fdf*) can take two forms, similar to those for the suicide simulations. The first is a *constant-rate* food supply function, supplying 50 units of food per cycle. The second is a *periodic* food supply function, which supplies food at the same constant rate, but is punctuated by severe droughts during which no food at all is generated. For this second *fdf*, the drought is exactly 8 cycles long,

Parameter	Value(s)
Gestation cycles	5
Food distribution function	Constant-rate — 50 food units per cycle Periodic drought — 50 food units for 40 cycles; 0 food units for 8 cycles
Actions and initial probabilities	Eat — $\frac{4}{9}$, Mate — $\frac{4}{9}$, Abortion — $\frac{1}{9}$
Observations	Health, Global food density, Is gestating?
Board size	25x25
Maximum entities per cell	1
Neighborhood	7x7
Initial population	400
Agent age limit	$N(100, 15^2)$ cycles
Health obtainable from food	$N(140, 10^2)$ health units
Health required for reproduction	200 health units
Genome type	Decision tree (variable structure) Decision tree (fixed structure; details in §6.5.9)
Action probability mutation	Initial mutation rates: $N(0, N(0.01, 0.001^2)^2)$, Meta-mutation rate: $N(0, 0.001^2)$
Branch node value mutation	Initial mutation rates: $N(0, N(0.01, 0.001^2)^2)$, Meta-mutation rate: $N(0, 0.001^2)$
Misc.	Male and Female Sexes Gestational investment uniform

Table 6.7: Parameters of the abortion simulations.

CHAPTER 6. EXPERIMENTS IN ETHICS

ACTION	SUCCESS		FAILURE	
	Utility	Health	Utility	Health
EAT	10	~70	-10	-10
ABORTION	0	-40	0	-40

MATE ACTION			
Outcome	Sex	Utility	Health
No conception		15	-16
Conception	F	15	-15
+ Total Gestation	F	0	-20
			-150
			-300
+ After Birth	F	0	-20
			-150
			-300
Conception	M	15	-10
Request denied		0	0
Cannot find mate		-10	-10

Table 6.8: The utilities and health effects associated with the outcomes of actions.

and there are 40 cycles between each drought. (Droughts arrive abruptly enough that agents will certainly notice them within the 5 cycles of a pregnancy.) The purpose of these two *fdfs* is to examine the effect of unexpected detrimental environmental changes on the rate of abortion.

Fixed Parameters

Gestation Period. During conception genetic material from both male and female parents is used to create a new agent. The female then carries this unborn agent — that is, she gestates — for a period of 5 cycles, unless she aborts. Gestation will also be cut short if the female dies, which will happen if the female's health drops to 0 or she reaches her maximum age.

Observations. Again to simplify, we allowed just three observations: self-health, global-food-density and is-gestating. The is-gestating observation is the only new one, allowing abortion attempts to be directed.

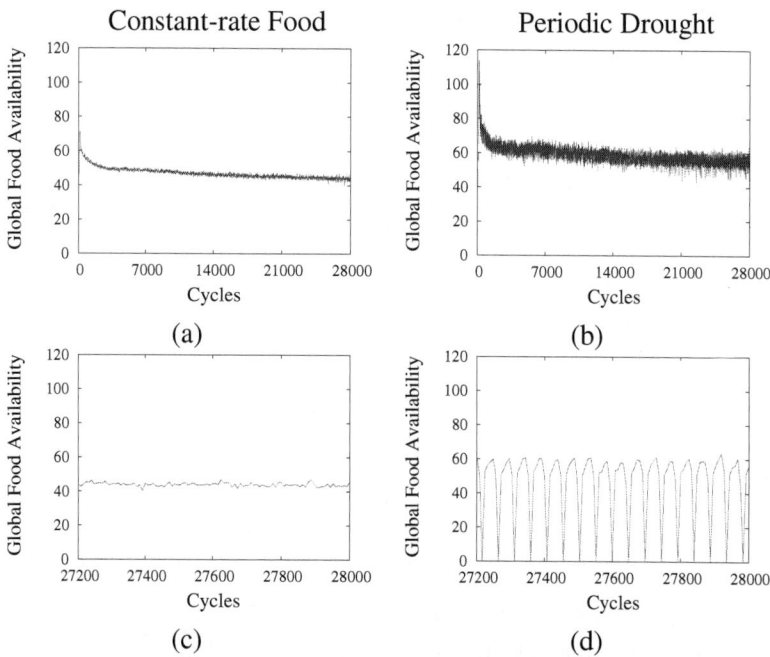

Figure 6.11: The amount of food available at the start of each cycle for (a) constant-rate food and (b) periodic drought experiments. Graphs (c) and (d) show the same experiments as (a) and (b), respectively, but restricted to the last 800 cycles.

Health and Utility. Table 6.8 presents the health and utilities for the various actions and outcomes. For successful matings, the negative values indicate the levels of investment — that is, the amount the parent loses and the amount the offspring gains.

Decision Function (Decision Trees). The decision trees are employed as described in detail in §4.6.2; both their structures and action distributions are evolvable. In some of the experiments here, however, we kept the structure of the decision tree fixed and only allowed action distributions to evolve. In this way, we could explore hypotheses about what agents will evolve under specific circumstances specified by the tree structure. For example, we could set the root branch node to test the gestation observation and thus easily separate the evolved strategies of gestating and non-gestating agents.

6.5.4 Basic Demographics and Orientation

Figure 6.11 shows the total amount of food available at the start of each cycle for constant-rate food and periodic drought simulations.[11] Figures 6.11b and 6.11d expand the last 800 cycles of each run set. Clearly, the constant-food simulations are relatively benign, while the cyclical-food simulations exhibit dramatic changes that are harder to accommodate. Figure 6.12 shows graphs for population, age and health from four run sets: one with constant-rate food and $abi = 300$, the other three with periodic drought and three levels of abi ($abi = 20, 150, 300$). Figures 6.12b-d clearly show the expected effect of periodic droughts on the agent populations, namely large variation in population size. The run set that produces the highest populations is the $abi = 20$ periodic drought run set (Figure 6.12b). Comparing this to Figures 6.12c and d, we can see that this is entirely due to the low abi. It is probable that if we allowed abi to evolve, it would evolve to this kind of lower level. Parental investment theory tells us that there needs to be a balance between starting an offspring's life with a better chance of surviving to reproductive age, and being able to have more offspring. This balance is struck at lower abi in our simulation, perhaps because food in the environment requires little effort to get. In any case, we explore the effects of a range of abi values, particularly higher values since they have the more pronounced effects and are closer to cases of interest.

Figures 6.12j-l show that average health is lower in simulations with lower abi. We might expect this, but low abi does not guarantee low average health. There is enough food in the environment for agents to build their health to higher levels. However, the population pressure seen in Figure 6.12b clearly stops agents from being able to eat more. Also note that average health when $abi = 150$ is much lower than when $abi = 300$, and sometimes it even dips below the health levels given $abi = 20$.

The average ages for periodic simulations (Figures 6.12f-h) are mostly unremarkable. While droughts do cause greater variation in the average age, this variation is muted in comparison to what happens with population and health in those same runs.

We have just seen that varying the *fdf* and *abi* parameters produces significant changes in some of the demographic statistics. We now explore the effect of these two parameters on abortion.

---

[11] All graphs here are taken from run sets of 20 simulations each unless otherwise noted.

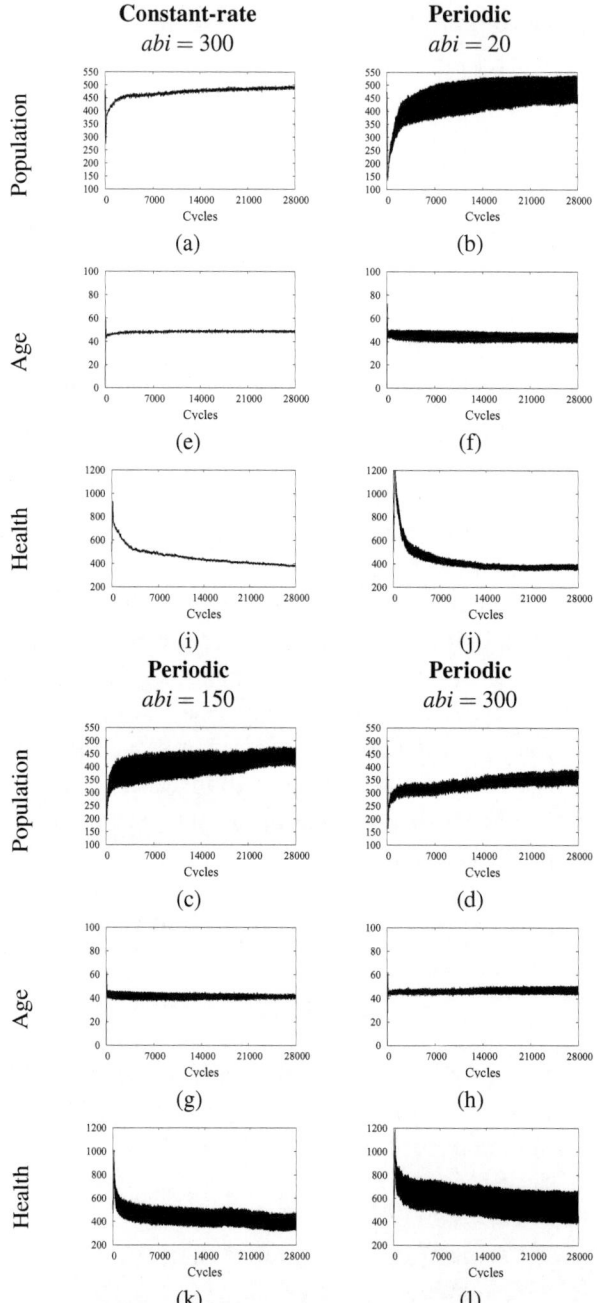

Figure 6.12: Population (a-d), age (e-h) and health (i-l) over four experiments: constant-rate food with $abi = 300$ and periodic food with $abi = 20, 150, 300$. All simulations here have no gestational investment.

CHAPTER 6. EXPERIMENTS IN ETHICS

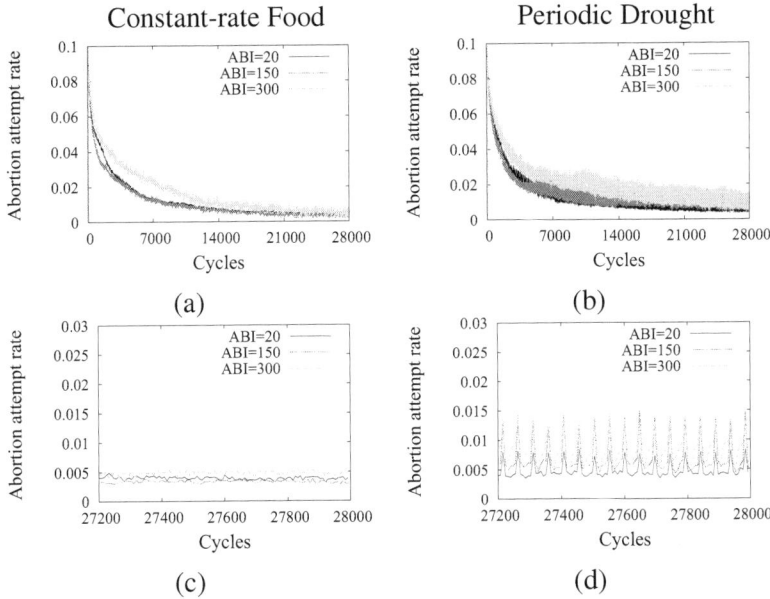

Figure 6.13: Abortion (attempt) rates in (a) constant-rate food and (b) periodic drought simulations. Each graph shows the rates for three levels of *abi* (when $gi = 0$). Graphs (c) and (d) display the last 800 cycles for these simulations.

6.5.5 Experiment: Varying After-birth Investment

There are two questions here: How do abortion rates change with changes in after-birth investment (*abi*)? And what effect does an unstable environment — that is, one with periodic droughts rather than a constant food supply — have on abortion?

We begin with Figure 6.13. Figures 6.13a and b show the abortion rates over time, given no gestational investment, for the constant-rate food and periodic drought experiments, respectively. Figures 6.13c and 6.13d are close-up views of the last 800 cycles of these runs.

Figures 6.13a and 6.13c show that over the long run *abi* does not correlate with abortion attempt rates (simply "abortion rates" subsequently) when there is a constant food supply. Nevertheless, with *abi* = 300, abortion rates evolve more slowly to their low final levels than in other cases. This suggests abortion does less damage to an agent's fitness when *abi* = 300 compared to when *abi* = 150 or *abi* = 20. However, it is clear that abortion is not adaptive in these cases.

By contrast, Figures 6.13b and 6.13d show that there is a positive correlation between *abi* and abortion rates when periodic droughts occur. The plots for $abi = 150$ and $abi = 300$ overlap, although a statistical significance test over the last 800 cycles shows the rates are distinct ($t(100) = 9.97$, $p \ll 0.001$), with the $abi = 300$ plot higher. While the difference between the $abi = 150$ and $abi = 300$ experiments is small, there is a clear difference between these and the case of $abi = 20$. In general, assuming some adaptive value to abortion, we should expect a positive correlation between abortion rates and *abi*, since higher *abi* means an abortion will save more potential investment for future reproduction. However, the average health of agents will also have a role to play. The average health of agents in the $abi = 150$ run set is, as noted earlier, always lower than that in the $abi = 300$ runs (and sometimes lower than the $abi = 20$ simulations). That implies that a saving of a health unit in the $abi = 150$ experiments will have a stronger effect on fitness than saving a unit in $abi = 300$ experiments. It is probably this difference that prevents a stronger correlation showing up between *abi* and abortion rates in the graph. We will see further evidence of this effect in the *gi* experiments of the next section.

From Figure 6.13d we can determine when the peaks in abortion rates occur. Unsurprisingly, they occur 48 cycles apart, matching the drought cycle. What is harder to see from the graphs, but established by examining the statistics, is that the peaks coincide exactly with the droughts themselves. In particular, the correlation coefficient between food generated and the abortion rate is -0.89 for the last 12000 cycles of the $abi = 300$ run set.

Interestingly, abortion rates begin to increase *before* a drought hits. After a drought, the population is at its smallest. As numbers recover, the world becomes ever more crowded, and it becomes increasingly difficult to find cells in which to put food — on average, only 43 food units per cycle can be generated during the most populated epochs, as opposed to the normal 50 units per cycle. In addition, the growing population consumes more of the available food. It is probable that these pressures make abortion a more beneficial option; agents endure lower health at these times, and are therefore more susceptible to being selected against, in both reproduction and survival. In consequence, even if we omit periods of drought, the negative correlation between food supply and abortion rates remains clear (-0.49).

In the constant-rate food simulations, there is continuous pressure on the food supply, yet abortion apparently has little adaptive value. The key difference is that an agent's environment in the constant-rate food simula-

CHAPTER 6. EXPERIMENTS IN ETHICS

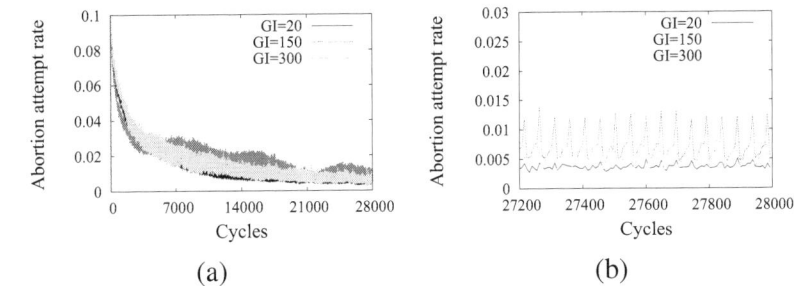

Figure 6.14: (a) Abortion rates in periodic drought experiments, with three levels of *gi* (and *abi* = 0). Graph (b) shows the last 800 cycles.

tions is completely *predictable*. Abortion is only of value in *partially predictable* environments, such as environments that are normally favorable, but that can occasionally show dramatic changes, as with droughts. Such changes can make it better in fitness terms to end a pregnancy rather than continue it.

6.5.6 Experiment: Varying Gestational Investment

Now we look at whether these results from varying *abi* also generalize to gestational investment (*gi*). We will not report on the constant-rate food simulations in this case, as their results were qualitatively the same as those using *abi*.

Figure 6.14a shows abortion rates from experiments with three levels of *gi*. We can readily note differences from the corresponding graph for *abi* (Figure 6.13b). Here, abortion rates do not evolve to be as robust. There is a great deal more variability in the way the simulations develop, with the $gi = 150$ experiments showing the strongest abortion rates.

The expanded view of the last 800 cycles (Figure 6.14b) is more revealing. It appears to show that when $gi = 20$ abortion rates are almost uncorrelated to droughts. The statistics concur, although the correlation between food generated and abortion rates is still present, at the weak level of -0.19. As the *gi* grows, the effect of droughts on abortion becomes stronger, and at $gi = 150$ the correlation is -0.64. However, when we reach the level of investment $gi = 300$, abortion rates are no longer correlated with droughts ($r = -0.09$). Instead, abortion rates are strongly correlated with the amount of food generated in the *preceding* epoch (-0.50). Indeed, the highest abortion rates in these runs occur just prior to droughts, not during the drought itself. Abortion rates remain moderately high during droughts, and reach their lowest levels immediately after, when food is most plentiful. These variations between $gi = 150$ and $gi = 300$ are clearly visible in Figure 6.14b.

The correlation between abortion rates and food generated in the preceding epoch for the $gi = 300$ run set is not as curious as it might seem. As with the *abi* experiments, the average health for $gi = 300$ is much higher than that for $gi = 150$ (the graphs are very similar to those in Figure 6.12). We already saw that the higher average health diminished the fitness of abortion somewhat in the $abi = 300$ simulations. With $gi = 300$, this factor significantly reduces the fitness value of abortion during droughts.

There are two points that suggest gestational investment is less important than after-birth investment in the evolution of abortion. First, an abortion will save an agent *all* of the *abi* it otherwise would have invested in its child, but on average only half of the *gi*. Second, an abortion during gestation will have wasted the *gi* made to that point. And we are *not* here falling for the Concorde fallacy — the agent will choose to abort if that choice gives the best expected *future* return to fitness. However, any gestational investment already made may tip the fitness balance in favor of continuing to birth, compared to aborting; this can never happen with an as yet unmade after-birth investment. Thus, it seems likely that parental investment in the form of *gi* is less likely to encourage abortion than when it is in the form of *abi*. We will see still further confirmation of this point shortly.

6.5.7 The Evolutionary Stability of Abortion

These simulations show that abortion can be an ESS, that is, its fitness is shown by an enduring tendency to use abortion as a part of a mixed reproductive strategy, a tendency well above the background rate of its reintroduction through mutation. For example, abortion in the $abi = 300$ runs occurs at a far higher rate than in the $abi = 20$ simulations. Indeed, even more strikingly, we can see that in these experiments, the abortion rate is systematically higher around droughts and lower otherwise. Clearly, abortion has been shaped by evolution in these simulations.

This outcome was not obvious a priori, for several reasons. Abortion causes an immediate reduction in an agent's reproductive success. It also throws away any gametic and gestational investment already put into reproduction. So, the evolutionary stability of abortion in the simulation, the positive fitness value of abortion, must be due to the unanticipated changes of conditions in the agent's world. Conditions may change for the worse, such as when food becomes rare, giving the female and any offspring a lower chance of surviving; or conditions may change for the better, such as

CHAPTER 6. EXPERIMENTS IN ETHICS

when food becomes plentiful, and the female would do better to take the opportunity to build up its health.

6.5.8 The Genetic Causes of Abortion

We now turn to an examination of the agent genomes to determine the conditions under which abortion is being chosen. As described in §6.5.3, the agent genome is a decision tree, in which branch nodes split on observables such as self-health, and leaves contain action distributions. Because of their complexity, it is difficult to calculate statistics on these genomes directly.

Figure 6.15 presents a table of the distributions of actual agent observations corresponding to different abortion rates, in experiments with $abi = 300$ and periodic droughts. The graphs omit observations made by non-gestating agents — other than the 'is-gestating' observation column, of course. Thus, we can, for example, compare the distributions of observed health for high and low abortion probabilities.

Figure 6.15 shows a clear relationship between the abortion probability and gestation, as we would expect, since without gestation abortion is simply self-harm. However, we would actually expect abortion to occur *solely* amongst gestating agents, which is not the case. There are still many non-gestating agents in the higher abortion probability ranges. Clearly, the agents have not been able to evolve decision trees capable of completely sorting out actual abortions from self-harm, even if the graph also demonstrates that they have made some progress. In the next section (§6.5.9), we reduce the evolutionary design space and will find that their behavior converges quite strongly as a result.

The food density column shows a swing to low food densities when we move from the rate 0-0.05 to the next higher ranges of abortion probability, as we would expect. Abortion rates from 0-0.05 and 0.05-0.10 already account for the majority of abortions and so could explain the abortion rate graphs of Figure 6.13. As abortion becomes more probable, the food distribution swings back towards higher food densities, although far fewer agents have abortion probabilities at these levels, so the selective pressure is less strong.

As for health, the distributions of this observation appear to become multimodal beyond the lowest range of abortion probabilities, with a slight increase in abortion rates for those with lower health and, in general, higher variance in health observations for those with higher abortion rates.

In order to compare the genomes from periodic drought and constant food experiments we can look to Figure 6.16. The distributions of the ges-

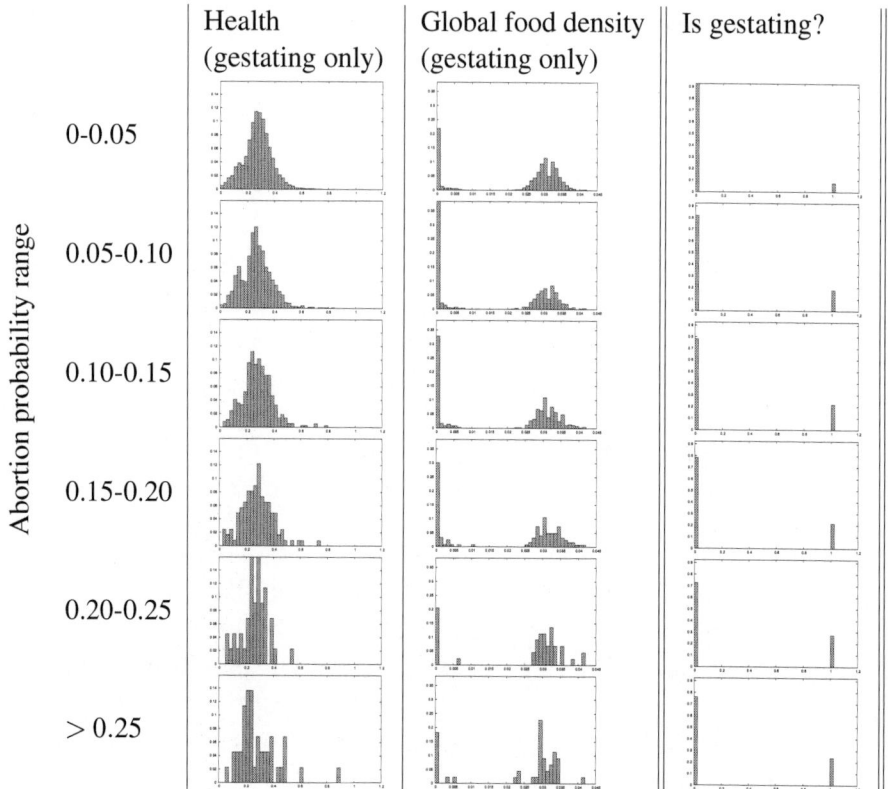

Figure 6.15: The distributions of observations for various ranges of abortion probabilities. Graphs are take from the periodic drought experiments, with $abi = 300$.

tation observations look similar to those in Figure 6.15, suggesting that abortion remains adaptive here. On the other hand, health observations vary less and do not as quickly break into multiple modes. Also, the food density observations show uniform normal distributions for the common cases (low abortion rates).

These graphs collectively suggest that abortion is more likely when agents have less food, lower health and are, in fact, pregnant. We can now look for clearer evidence for these conditions by introducing fixed genomes into the simulations, which will make the analysis much simpler.

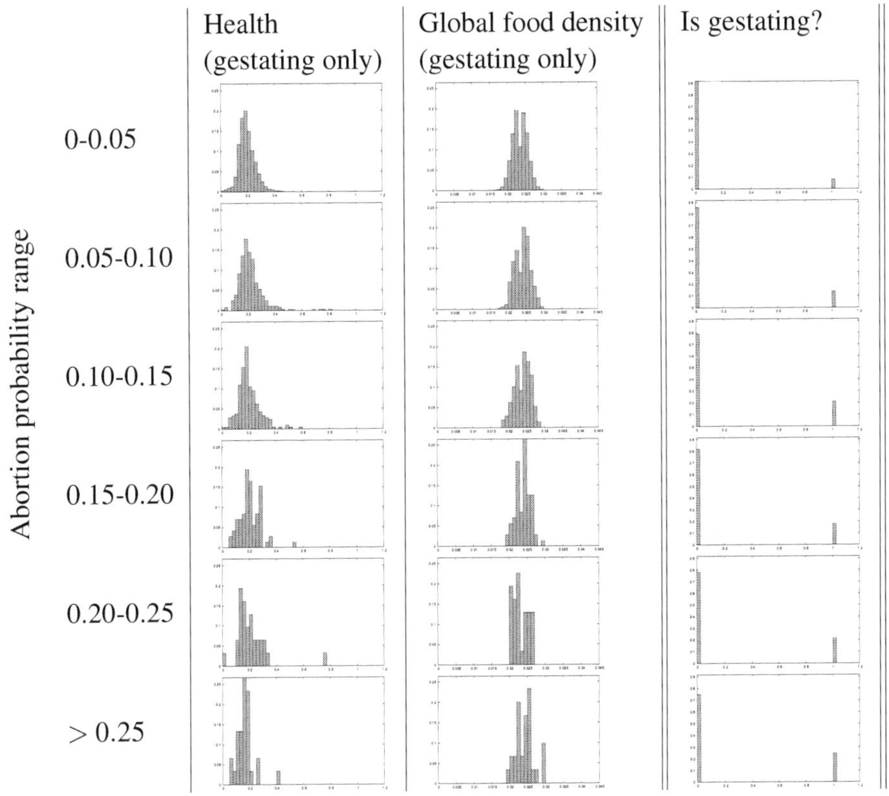

CHAPTER 6. EXPERIMENTS IN ETHICS

	Health (gestating only)	Global food density (gestating only)	Is gestating?
0-0.05			
0.05-0.10			
0.10-0.15			
0.15-0.20			
0.20-0.25			
> 0.25			

Abortion probability range

Figure 6.16: The distributions of each observation for various abortion rates, given constant food supply and $abi = 300$.

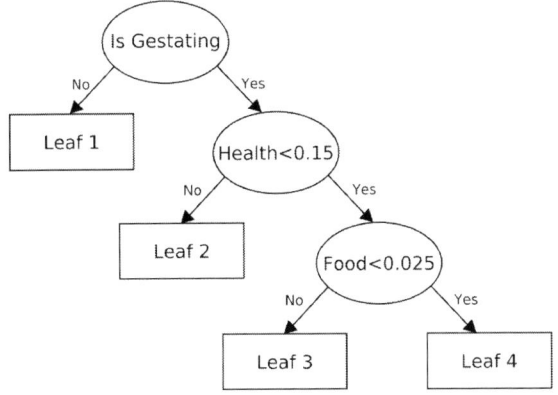

Figure 6.17: The fixed decision tree structure used; action probabilities in the leaves were free to evolve.

6.5.9 Introducing Fixed Genomes

Earlier we saw that abortion is strongly correlated with droughts and that there are correlations between abortion and the levels of *abi* and, to a lesser extent, of *gi*. Together, these correlations confirm the hypothesis that abortion can be part of an ESS when the environment can unexpectedly turn for the worse and the agent can use an abortion to conserve investment for more propitious times. To produce an even more convincing test of this explanation of abortion's fitness value, we fixed the structure of the agents' decision trees to reflect the hypothesis.

For the fixed decision tree each branch split on a relevant condition, with one path down the tree embodying all the conditions of the adaptation hypothesis. Then we simply let the action probabilities in the leaves evolve and observed the results. If the leaf on the hypothesis path — which describes evolved probabilities for circumstances matching the conjunction of all the conditions — evolves systematically higher abortion probabilities than those for the other paths, we have a confirmation of the adaptation hypothesis.

The fixed genome structure for this experiment is shown in Figure 6.17. The path corresponding to the adaptation hypothesis ends at Leaf 4; this leaf triggers when an agent is gestating, when it has low health (i.e., health < 0.15) and when the global food density is low. Of course, this does not describe our adaptation hypothesis exactly; that is, there is no node that asserts that there are "unexpected and difficult circumstances". However, these conditions constitute such circumstances for our simulation.

We described the recombination of fully evolvable decision trees in §4.6.2. Recombination for fixed-structure decision trees is the same, except that there are no mutations of branch nodes — that is, the tree structure, branch type and branch split-values do not change. In these experiments, parent decision trees always have the same structure, so the result of reproduction is a child genome with, again, identical structure, but containing mutated copies of parent leaves.

Figure 6.18 shows the rates of abortion attempts in this simulation for both constant-rate and periodic food experiments, which can be compared to Figure 6.13. We can see a few differences from the earlier simulations. In the periodic drought conditions (Figure 6.18b and 6.18d) the correlation between *abi* and abortion rates still appears, but it has grown much more significant. At $abi = 300$, abortion rates evolve to peak at much higher rates than before. In contrast, at $abi = 150$ abortion peaks at lower rates than in the corresponding prior experiments, while the $abi = 20$ experiments are

CHAPTER 6. EXPERIMENTS IN ETHICS

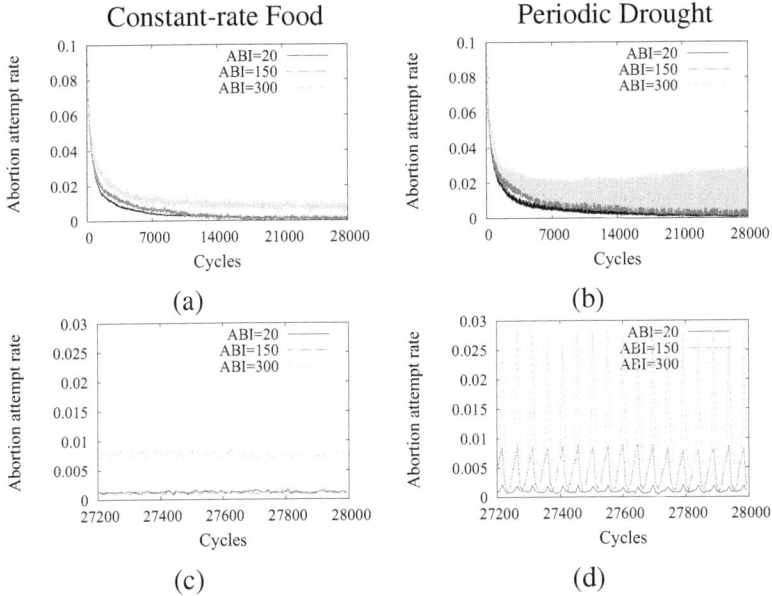

Figure 6.18: Abortion rates in (a) constant-rate food and (b) periodic drought experiments with fixed decision trees. Each graph shows the three levels of *abi* (with $gi = 0$). Graphs (c) and (d) show the last 800 cycles for experiments (a) and (b), respectively.

substantially more muted in their effects. Also of interest is the gap between trough and peak, which has grown for $abi = 300$, declined for $abi = 150$, and nearly vanished for $abi = 20$.

In the constant-rate food simulations, the $abi = 150$ and $abi = 20$ experiments look similar to before, though abortion has evolved to an even lower rate. In contrast, the $abi = 300$ simulations have evolved very differently, with the abortion rate stabilizing much higher. Earlier in §6.5.5, we noted the slower decline of abortion rates when $abi = 300$. Here, we see this slight positive value for abortion asserting itself more forcefully. It seems that seeding clear-cut conditions in the genome that are favorable for abortion has allowed abortion to persist in a world with constant food.

We also conducted simulations varying *gi* rather than *abi*, with Figure 6.19 showing the results. We can see a similar pattern, but with much reduced effect sizes. In particular, the simulations with $gi = 20$ and $gi = 150$ do not show statistically significant differences between the abortion attempt rates and the background mutation rate.

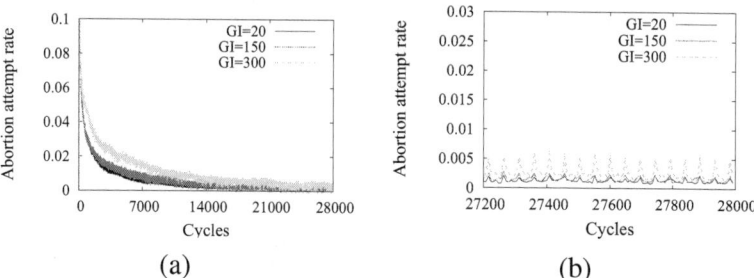

Figure 6.19: (a) Abortion rates in periodic drought runs with fixed tree structures, showing three levels of *gi* (with *abi* = 0). Graph (b) shows the last 800 cycles.

	Eat		Mate		Abort	
Condition	Mean	SD	Mean	SD	Mean	SD
Not Gestating	0.5958	0.0014	0.4030	0.0014	0.0012	0.0000
Gestating, Health$\geq$.15	0.9812	0.0011	0.0161	0.0010	0.0027	0.0001
Gestating, Health$<$.15, Food$\geq$.025	0.8976	0.0053	0.0136	0.0007	0.0888	0.0050
Gestating, Health$<$.15, Food$<$.025	0.2377	0.0402	0.3671	0.0546	0.3952	0.0258

Table 6.9: Average evolved action probabilities for each of the conditions in the fixed decision tree, along with standard deviations.

Examining the Genetic Causes

Figure 6.20 shows the periodic drought experimental set with *abi* = 300 in the same arrangement as Figure 6.15. The figure is quite striking. Whenever abortion probabilities are high, agents are gestating — as we speculated should occur before. Clearly, the problem of finding this adaptive solution has been greatly simplified for evolution. Also, higher abortion probabilities have become very strongly correlated with low health. Finally, in the lower abortion probability ranges global food densities are mostly clustered around higher densities. But as the abortion probability increases, abortion rates are strongly associated with low food density.

The fixed genome structure also allows us to get an excellent idea of what the typical genome, in particular its action probabilities, looks like. Table 6.9 shows the the mean action probabilities for each leaf of the fixed decision tree of Figure 6.17 for the last quarter of the epochs sampled for periodic drought simulations with *abi*=300, along with their standard deviations.

As we would expect, the abortion rate rises in step with the number of conditions of the adaptation hypothesis being matched. At Leaf 1 (the

CHAPTER 6. EXPERIMENTS IN ETHICS 227

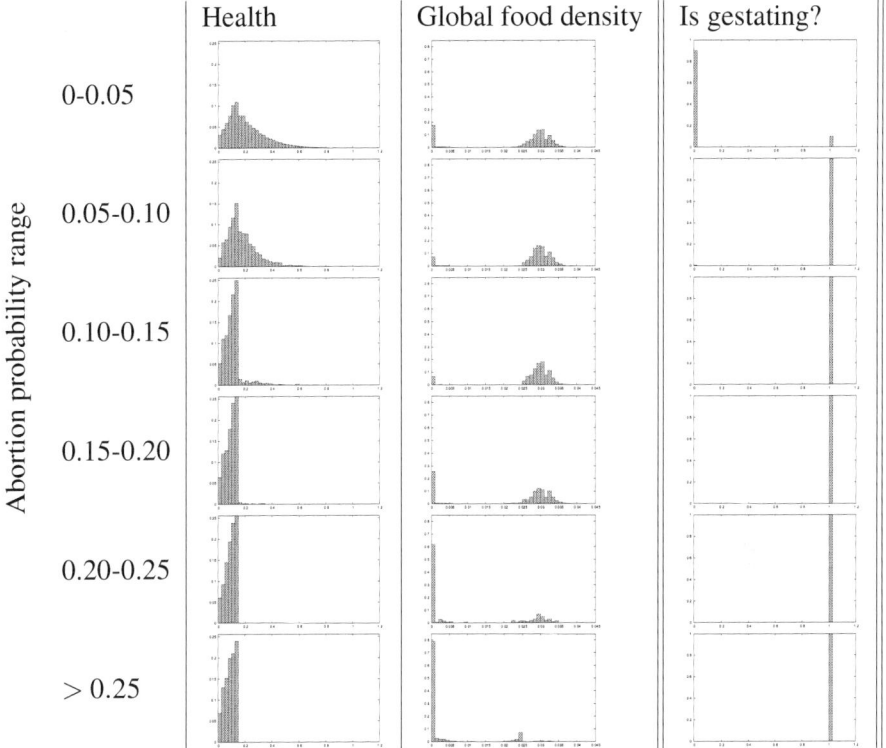

Figure 6.20: The distributions of each observation for various ranges of abortion probabilities.

first row), we can see that abortion has all but disappeared, as we would expect, since it is simply self-harm. At Leaf 2, the probability of abortion is slightly higher. This leaf only covers high health conditions, but it does not discriminate between low and high food densities. At Leaf 3 — where health is low, but food density is high — the probability of abortion grows significantly greater. Finally, at Leaf 4, where the agent is gestating, health is low and food density is low, the abortion probability is very high. It does not have probability near 1, so selection pressure is not absolute in these conditions, but a mixed strategy incorporating abortion is clearly adaptive. We can also see that mating has a substantially higher probability than eating, which might be surprising given that the agent is pregnant; but this is simply a case of the agents evolving a trade-off between mating and eating to get the highest health payoff — or, rather, the least worst payoff. The reason for this is that the net health effect of mating is slightly less harmful than that of eating, *under these conditions* — i.e., where mating cannot produce another offspring *and* the agent is quite unlikely to find any food to eat.

The conclusion from these experiments is unambiguous: in the circumstances simulated here, abortion has adaptive value for dealing with unexpectedly difficult times, when the agent can save substantial health for future parental investments.

6.5.10 Experiment: Exploring the Ethics of Abortion

Unlike suicide and rape, the ethics of abortion is a highly contentious *political* issue. As we noted earlier, we cannot here explore the single most controversial issue involving abortion: the ethical status of the fetus. And even limiting ourselves to consequentialist implications, the simulation is currently too limited to investigate many issues of interest. Nevertheless, our simulations still yield interesting results that are at least suggestive of when and how abortion can be the more ethical choice.

Figure 6.21 shows the cumulative utilities from several experiments varying values for after-birth investment, excepting Figure 6.21e which describes experiments varying gestational investment instead.[12] We can see from these graphs that *abi* is negatively correlated with cumulative total utility, regardless of whether there is any abortion in the simulation. The connection between lower *abi* and higher utility is almost certainly due to the higher populations that lower *abi* allows (compare Figures 6.12b, 6.12c and 6.12d). This same relation holds when the parental investment is in the form of *gi* rather than *abi* (Figure 6.21e).

If we compare the constant-rate food with periodic drought experiments (i.e., the first with the second row) droughts seem to have a mixed effect on cumulative utility. Figure 6.22 pairs the constant-rate food and periodic drought experiments across various levels of *abi* for simulations that do and do not contain abortion. Clearly, as the *abi* increases (moving top to bottom in the figure), the gap between the utilities for periodic droughts versus constant-rate food grows; that is, with greater after-birth investment the negative utilitarian impact of drought increases. Comparing experiments with abortion (second column) with those without (first column), there does not seem much difference. However, when $abi \geq 150$, the availability of abortion reduces the impact of drought somewhat.

---

[12]The graphs for this section are all taken from the experiments from §§6.5.4-6.5.8, which contain evolvable genomes.

CHAPTER 6. EXPERIMENTS IN ETHICS

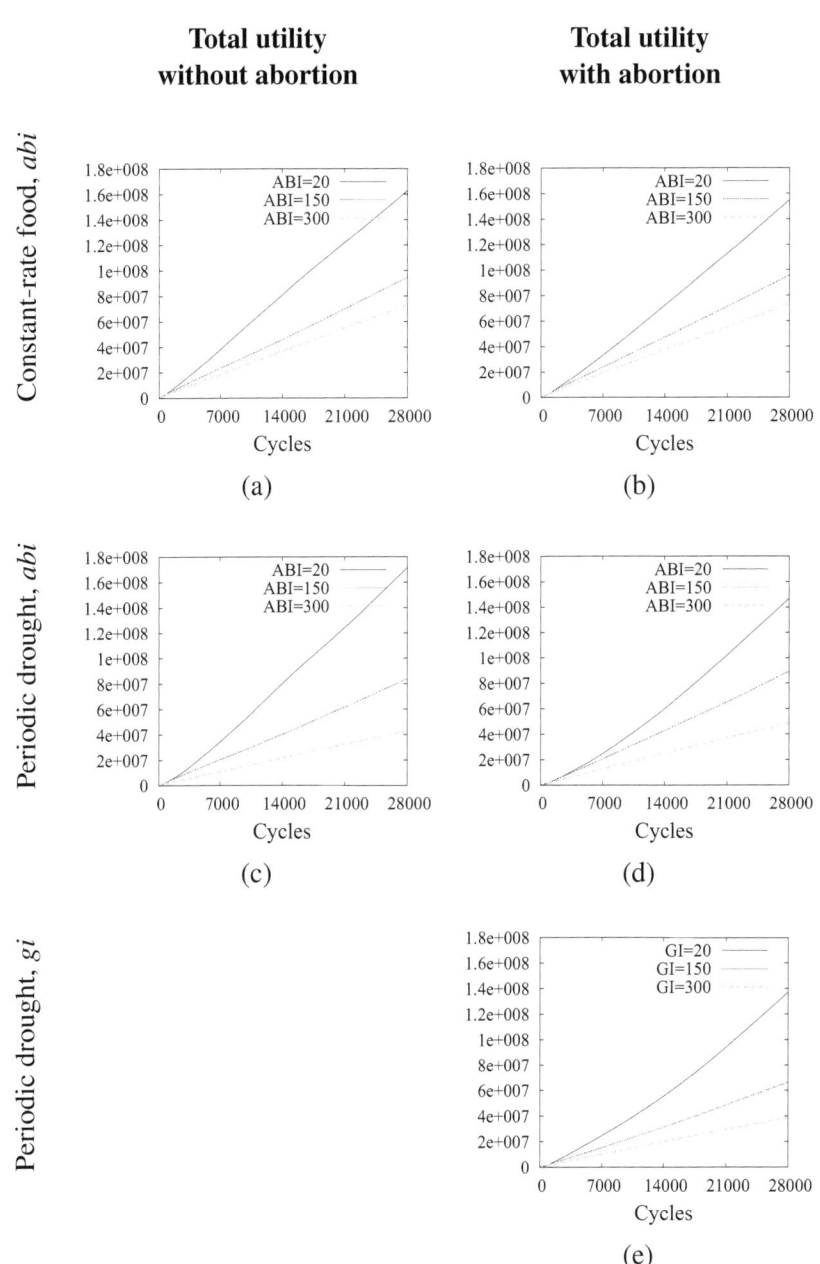

Figure 6.21: The cumulative total utilities in experiments with (a-b) constant food supply and (c-e) periodic drought, with (right column) and without abortion (left column).

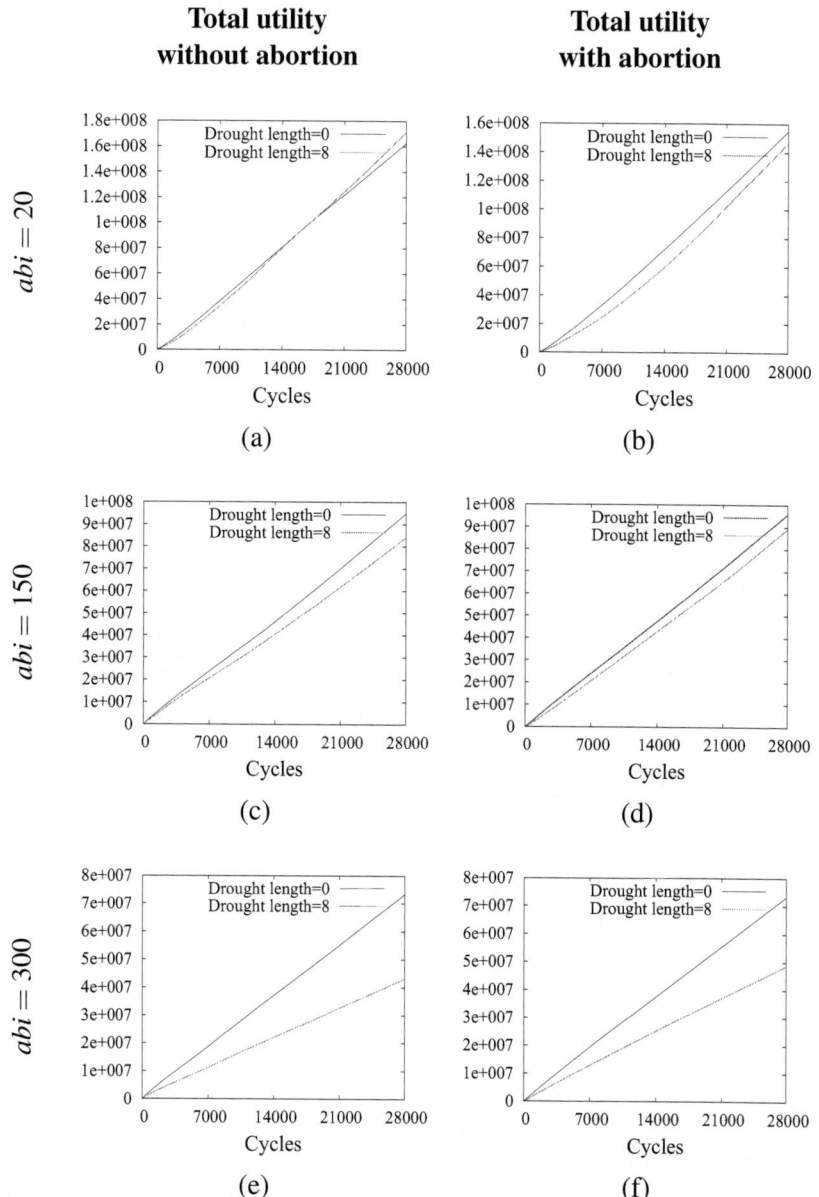

Figure 6.22: The cumulative utilities in experiments without abortion (a, c and e) and matched experiments with abortion (b, d and f), with varying *abi*. In order to highlight the differences, the vertical scales of each graph are *not* the same.

CHAPTER 6. EXPERIMENTS IN ETHICS

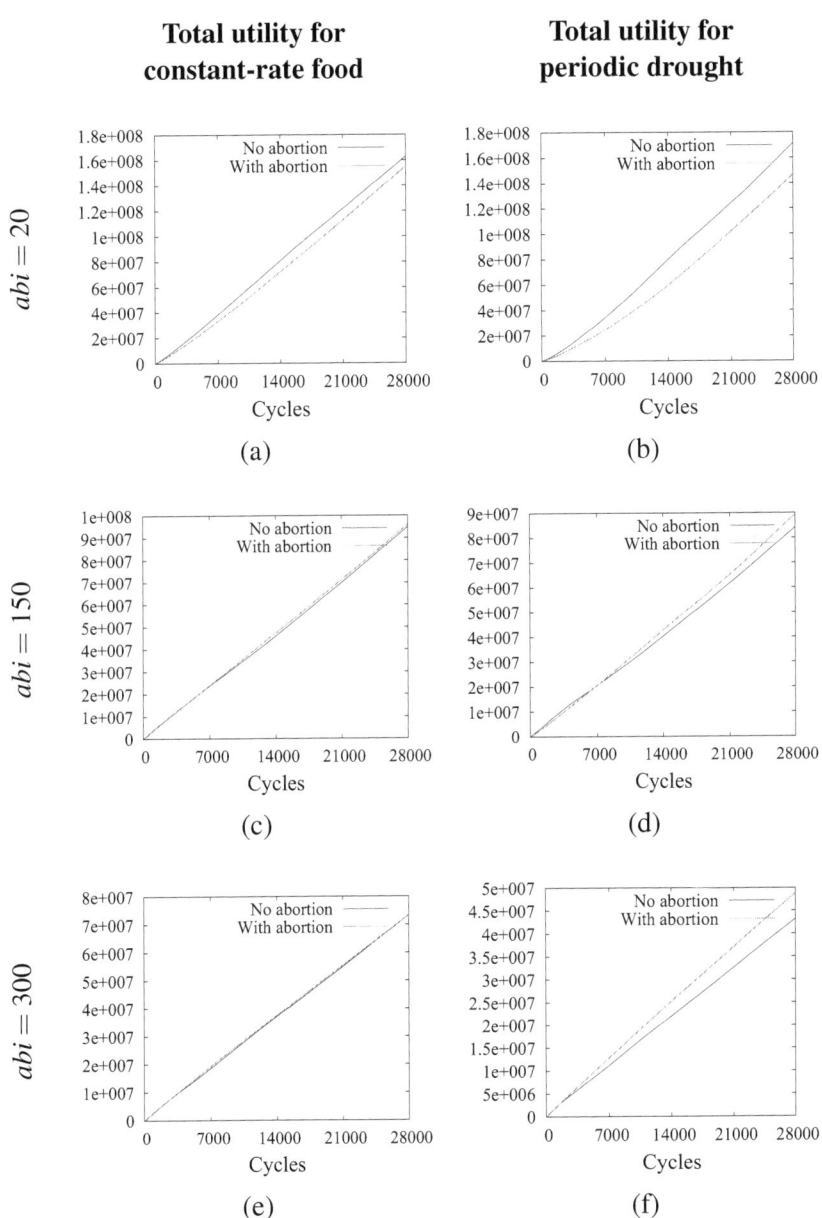

Figure 6.23: The cumulative utilities in experiments with constant food (a, c and e) and periodic drought (b, d and f), varying *abi*. In order to highlight differences, the vertical scales of each graph are *not* the same.

Figure 6.23 shows the same information in a different format, comparing matched experiments with and without abortion for each combination of food distribution (columns) and *abi* (rows). The first thing we can see is that abortion has no positive effect in the constant food simulations. Indeed, in the $abi = 20$ experiments, the effect of abortion on utility is clearly negative. In these simulations abortion also evolved away, down to the background rate of its reintroduction by mutation. This low level of abortion was sufficient to provide a constant drain on the total utility of the population.

The situation in the simulations containing droughts is quite different (Figures 6.23b, d and f). When $abi = 20$, abortion produces ethically unfavorable outcomes. It could be the case here that abortion has value for the agent that aborts, but disvalue for others, including the offspring itself; or, perhaps, the reintroduction rate is again imposing abortion on agents in suboptimal situations. As the *abi* increases, however, abortion's utilitarian benefit also increases. When $abi = 150$, abortion is moderately beneficial, while when $abi = 300$, abortion clearly allows for more ethical outcomes overall, from the utilitarian perspective.

6.5.11 Summary

There has been little discussion in the literature on how abortion may evolve. And most of that discussion has focused on the evolution of spontaneous abortion — mainly because induced abortion is apparently limited to humans and has only become safe and effective in recent times. However, some, such as Hrdy (1999) and Lycett and Dunbar (1999), have also suggested evolutionary paths for induced abortion.

The problem of when abortion will be adaptive still lingers. We saw that if circumstances can often change unexpectedly for the worse after conception, with droughts representing such changes in our experiments, then abortion can provide a fitness benefit, particularly if an offspring needs substantial after-birth investment. This hypothesis was supported by the correlation between droughts in food and peaks in abortions in our studies. The hypothesis of the adaptive value of abortion was supported further by experiments with fixed genome structures that tested the conditions of the adaptation hypothesis directly.

The experiments also showed a large difference in outcomes for parental investments occurring after birth (*abi*) and for parental investments occurring during gestation (*gi*). In the after-birth case, there was a positive correlation between the level of investment and the average abortion rate, which was particularly strong in the fixed genome structure experiment. In the gestational case, the connection between the level of investment and abortion was far weaker. This is clearly due to differences in the average amount of investment saved in each case and how much investment has already been made.

Finally, our experiments looking into the utilitarian implications of abortion produced mixed results. In constant food simulations, abortion has no effect or a minor negative effect. In periodic drought scenarios, however, the effects of abortion range from a negative effect when the amount of investment needed is low ($abi = 20$) to a positive effect when required investment is high ($abi = 150$ and $abi = 300$). In the latter cases, the utilitarian value of abortion corresponds to its adaptive value.

6.6 Conclusion

We believe agent-based models will become a preferred tool for investigating all varieties of evolved behavior, not just the behavior of crowds and markets of the social simulationists, but also the purposive ethical (and unethical) behavior of applied philosophy. We've chosen two cases, rape and abortion, to illustrate some of the potential here. It's clear that these simulation tools show good potential to test many of the hypotheses of ethology and evolutionary psychology that would otherwise be difficult or impossible to explore empirically. But equally they have the potential to cast light upon ethical claims — at the least, the ethical claims made by utilitarians.

Chapter 7

The Future

Our political representatives gather around a table, grave looks upon their faces. There is increasing pressure from both the public and the fourth estate to remove costly welfare support and to return the savings to the public in the form of tax cuts. Economic models show that instituting such a policy would boost both total and per capita GDP over the long term. Sociologists argue based on historical anecdotes and intuition that such policies will cause short term harm but long term gain. Nonetheless, the politicians feel uneasy. Believing they lack crucial information, they commission a series of utilitarian simulation studies to explore the effect on the nation's well-being. After a year, the results are in. While the specific results vary, one clear phenomenon emerges from every simulation: such a policy substantially harms not only short-term utilities but long-term cumulative utilities as well.

Ethical simulations will not be playing such a focused role in policy making any time soon. When faced with difficult policy choices, a broader public has yet to accept that simulations can provide relevant evidence. In the debates over anthropogenic climate change, simulation models have made a convenient punching bag for skeptics and deniers (Norton and Suppe, 2001). Despite our efforts here, and those of fellow simulationists, wide public acceptance of simulation's capacity to inform political matters may be some way off.

The concern is not entirely unfounded. Simulations capable of modeling very specific policies within the political, economic and social framework of a given country or the world as a whole are difficult to do well with current technology, and especially with current software technology, despite the enormous progress that has been made since the very early and simple attempts of the Club of Rome (Meadows, 1972). Still, there are many impressive simulations in use today in economics, epidemiology, and, not least, climatology. Indeed, the "economic models" pitted against our ethical simulations in our brief fantasy above can only have been telling economists and politicians a story revealed through computer simulations, however incompletely designed. And perhaps that suggests the most plausible route that ethical simulation might find a political hearing: economic simulation is meant to be the simulation of the body politic's well being, and so it is very naturally enlarged to include the explicit representation of its members' quality of life.

While the benefits of ethical simulation may take some time to flow through to political decision making, philosophy can enjoy the benefits today. There is a legion of social behaviors and constructs that have remained problematic: euthansia, murder, theft, lying, generosity, corporations, forms of government and social welfare, to name a few. Our interest in these stems beyond the simple question, When is it ethical? We also want to know *why* and *how* they might be ethical — as examples, how does a behavior impact on the behavior of others; what institutions combine with what behaviors to promote positive or negative outcomes; and how does a behavior or social institution come to exist and remain stable at all?

Here we have examined a very small sample of phenomena of interest to ethicists. Along the way, we discovered: how altruism can flourish, reinforcing empirically the theoretically established claims for inclusive fitness and group selection pressure; how rapid aging can promote a diversity response to co-evolutionary (or other) health challenges to a population; how anatomical sexes can evolve sexually dimorphic behaviors; how rape, even when divorced of specific negative consequences, remains firmly unethical; and how abortion can both evolve and be ethical when the environmental costs of continuing a pregnancy are too high. Undeniably our simulations here are simple in design, even if fairly complex in detail. Perhaps some will complain that they are simply naive. But if they are, and to the extent that they are, it is not because of the nature of the enterprise we have undertaken, but either because of our overlooking important elements in our designs or because of our inability to simulate the levels of detail and complexity required to reveal yet more surprising and important emergent behavior. In either case, the onus is rightly put upon the objector: go forth and simulate better!

Bibliography

Adami, C. (1998). *Introduction to Artificial Life*. New York: Springer.

Agustí, S., M. Satta, M. Mura, and E. Benavent (1998). Dissolved esterase activity as a tracer of phytoplankton lysis: Evidence of high phytoplankton lysis rates in the northwestern mediterranean. *Limnology and Oceanography 43*, 1836–1849.

Al-Ansary, L. A., G. Oni, and Z. A. Babay (1995). Risk factors for spontaneous abortion among saudi women. *Journal of Community Health 20*(6), 491–500.

Alexander, J. and B. Skyrms (1999). Bargaining with neighbors: is justice contagious? *The Journal of Philosophy 96*, 588–598.

Anscombe, G. E. M. (1958). Modern moral philosophy. *Philosophy 33*, 1–19.

Appalanaidu, C. (2007). Group selection: An investigation into the potential for the evolution of virtual ecosystems. Clayton School of IT, Monash University. Honours Thesis.

Aristotle (325BC/1998). *The Nicomachean Ethics*. Oxford: Oxford Univ.

Asimov, I. (1954). *The Caves of Steel*. Garden City, NY: Doubleday.

Axelrod, R. (1984). *The Evolution of Cooperation*. New York: Basic Books.

Axelrod, R. (1997). *The Complexity of Cooperation: Agent-Based Model of Competition and Collaboration*. Princeton, NJ: Princeton University Press.

Axelrod, R. and W. D. Hamilton (1981). The evolution of cooperation. *Science 211*, 1390–1396.

Axtell, R., R. Axelrod, J. M. Epstein, and M. D. Cohen (1996). Aligning simulation models: A case study and results. *Computational and Mathematical Organization Theory 1*, 123–141.

Baldwin, M. J. (1896). A new factor in evolution. *The American Naturalist 30*, 441–451.

Bateman, A. J. (1948). Intra-sexual selection in *Drosophila*. *Heredity 2*, 349–368.

Batterman, R. (2007). Intertheory relations in physics. *Stanford encyclopedia of philosophy*. http://plato.stanford.edu/.

Bedau, M. (2006). The evolution of complexity. *Symposium on the Making up of Organisms*, Ecole Normale Suprieure.

Belenky, P., F. Racette, K. Bogan, J. McClure, J. Smith, and C. Brenner (2007). Nicotinamide riboside promotes Sir2 silencing and extends lifespan via Nrk and Urh1/Pnp1/Meu1 pathways to NAD+. *Cell 129*, 473–484.

Bentham, J. (1789). *An Introduction to the Principles of Morals and Legislation*. Oxford: Clarendon Press (1996). Edited by J. H. Burns and H. L. A. Hart.

Berlekamp, E., J. Conway, and R. Guy (1982). *Winning Ways for Your Mathematical Plays*, Volume 2. New York: Academic Press.

Bishop, M. and J. D. Trout (2005). *Epistemology and the Psychology of Human Judgment*. Oxford: Oxford University Press.

Borges, J. (1954). *Historia Universal de la Infamia*. Buenos Aires: Emece.

Bostrom, N. (2009). The infinitarian challenge to aggregative ethics. http://www.nickbostrom.com/ethics/infinite.pdf.

Boyd, R., H. Gintis, S. Bowles, and P. Richerson (2003). The evolution of altruistic punishment. *Proceedings of the National Academy of Sciences 100*(6), 3531–3535.

Boyd, R. and P. Richerson (1992). Punishment allows the evolution of cooperation (or anything else) in sizable groups. *Ethology and sociobiology 13*(3), 171–195.

Bredesen, D. (2004). The non-existent aging program: How does it work. *Aging Cell 3*, 255–259.

Brinkers, M. and P. den Dulk (1999). The evolution of non-reciprocal altruism. In D. Floreano, J.-D. Nicoud, and F. Mondada (Eds.), *Advances in Artificial Life*, Volume 1674 of *Lecture Notes in Computer Science*, pp. 499–503. Berlin: Springer.

Brown, R. M., E. Dahlen, C. Mills, J. Rick, and A. Biblarz (1999). Evaluation of an evolutionary model of self-preservation and self-destruction. *Suicide and Life-Threatening Behavior 29*(1), 58–71.

Brownmiller, S. and B. Mehrhol (1992). A feminist response to rape as an adaptation in men. *Behavioral and Brain Sciences 15*(2), 381–382.

Bruce, H. M. (1960). A block to pregnancy in the house mouse caused by the proximity of strange males. *Journal of Reproduction and Fertility 1*, 96–103.

Buss, D. M. (1988). The evolution of human intrasexual competition: Tactics of mate attraction. *Journal of Personality and Social Psychology 54*, 616–628.

Buss, D. M. (1999). *Evolutionary Psychology: The New Science of the Mind*. Boston: Allyn and Bacon.

Buss, D. M., R. J. Larsen, D. Westen, and J. Semmelroth (1992). Sex differences in jealousy: Evolution, physiology, and psychology. *Psychological Science 3*, 251–255.

Chalmers, A. F. (1982). *What is This Thing Called Science?* St. Lucia, Qld: University of Queensland Press.

Chen, W.-H., X.-X. Wang, W. Lin, X.-W. He, Z.-Q. Wu, Y. Lin, S.-N. Hu, and X.-N. Wang (2006). Analysis of 10,000 ESTs from lymphocytes of the cynomolgus monkey to improve our understanding of its immune system. *BMC Genomics 7*.

Cheng, P. and K. Holyoak (1985). Pragmatic reasoning schemas. *Cognitive Psychology 17*(4), 391–416.

Chomsky, N. (1959). A review of B. F. Skinner's verbal behavior. *Language 35*(1), 26–58.

Clutton-Brock, T. (1991). *The Evolution of Parental Care*. Princeton, NJ: Princeton University Press.

Clutton-Brock, T. H. and G. A. Parker (1995). Sexual coercion in animal societies. *Animal Behavior 49*, 1345–1365.

Copeland, B. J. (2002). Hypercomputation. *Minds and Machines 12*, 461–502.

Cosmides, L. (1989). The logic of social exchange: Has natural selection shaped how humans reason? Studies with the wason selection task. *Cognition 31*, 187–276.

Cosmides, L. and J. Tooby (1987). From evolution to behavior: Evolutionary psychology as the missing link. In J. Dupré (Ed.), *The Latest on the Best: Essays on Evolution and Optimality*, pp. 277–306. Cambridge, MA: MIT Press.

Cosmides, L. and J. Tooby (2000). Consider the source: The evolution of adaptations for decopuling and metarepresentation. In D. Sperber (Ed.), *Metarepresentations: A multidisciplinary approach*, pp. 53–115. New York: Oxford University Press.

Coyne, J. A. (2003). Of vice and men: A case study in evolutionary psychology. In C. B. Travis (Ed.), *Evolution, Gender, and Rape*, pp. 171–190. Cambridge, MA: MIT Press.

Daly, M. and M. Wilson (1982). Whom are newborn babies said to resemble? *Ethology and Sociobiology 3*, 69–78.

Daly, M. and M. Wilson (1994). Some differential attributes of lethal assaults on small children by stepfathers versus genetic fathers. *Ethology and Sociobiology 15*, 207–217.

Darwin, C. (1880). *Descent of Man, and Selection in Relation to Sex*. New York: D. Appleton & Co.

Darwin, C. (1988/1859). *On the Origin of Species*. Washington Square, NY: New York University Press.

Dawkins, R. (1976). *The Selfish Gene*. Oxford: Oxford University Press.

Dawkins, R. and T. Carlisle (1976). Parental investment, mate desertion and a fallacy. *Nature 262*, 131–133.

de Catanzaro, D. (1981). *Suicide and Self-damaging Behavior: A Sociobiological Perspective*. New York: Academic Press.

de Catanzaro, D. (1986). A mathematical model of evolutionary pressures regulating self-preservation and self-destruction. *Suicide and Life-Threatening Behavior 16*, 166–181.

de Catanzaro, D. (1995). Reproductive status, family interactions, and suicidal ideation: Surveys of the general public and high-risk groups. *Ethology & Sociobiology 16*, 385–394.

den Dulk, P. and M. Brinkers (2000). Evolution of altruism in viscous populations: effects of altruism on the evolution of migrating behavior. In *Parallel Problem Solving from Nature PPSN VI*, pp. 457–466. Springer.

Dennett, D. (1991). *Consciousness Explained*. Little, Brown & Co.

Dennett, D. (1995). *Darwin's Dangerous Idea*. New York: Simon and Schuster.

Devereux, G. (1954). A typological study of abortion in 350 primitive, ancient and pre-industrial societies. In H. Rosen (Ed.), *Therapeutic Abortion*. New York: The Julian Press Inc.

Di Paolo, E. A., J. Noble, and S. Bullock (2000). Simulation models as opaque thought experiments. In M. A. Bedau, J. S. McCaskill, N. H. Packard, and S. Rasmussen (Eds.), *Artificial Life VII: Proceedings of the Seventh International Conference on Artificial Life*, Cambridge, MA, pp. 497–506. MIT Press.

Diamond, J. (1997). *Why is Sex Fun?* Basic Books.

Dobzhansky, T. (1951). *Genetics and the Origin of Species*. New York: Columbia University.

Dorin, A. and K. Korb (2007). Building virtual ecosystems from artificial chemistry. In *Proceedings of the Ninth European Conference on Advances in Artificial Life*, pp. 103–112.

Dorin, A., K. Korb, and V. Grimm (2008). Artificial-life ecosystems: What are they and what could they become? In *The Eleventh International Conference on the Simulation and Synthesis of Living Systems*, pp. 173–180.

Dresher, M. (1961). *The Mathematics of Games of Strategy: Theory and Applications*. Englewood Cliffs, NJ: Prentice-Hall.

Dreyfus, H. L. (1992). *What Computers Still Can't Do: A Critique of Artificial Reason* (3rd ed.). Cambridge, Mass: MIT Press.

Dupré, J. (1987). *The Latest on the Best: Essays on Evolution and Optimality*. Cambridge, MA: MIT Press.

Eagly, A. H. and W. Wood (2003). The origins of sex differences in human behavior: Evolved dispositions versus social roles. In C. B. Travis (Ed.), *Evolution, Gender, and Rape*, pp. 261–304. Cambridge, MA: MIT Press.

Edmonds, B. and D. Hales (2005). Computational simulation as theoretical experiment. *Journal of Mathematical Sociology 29*, 209–232.

Eidelson, B. M. and I. Lustick (2004). Vir-pox: An agent-based analysis of smallpox preparedness and response policy. *Journal of Artificial Societies and Social Simulations 7*(3).

Eldredge, N. and S. J. Gould (1972). Punctuated equilibria: an alternative to phyletic gradualism. In T. J. M. Schopf (Ed.), *Models in Paleobiology*, San Francisco, pp. 82–115. Freeman, Cooper and Company.

Epstein, J. M. and R. Axtell (1996). *Growing Artificial Societies: Social Science from the Bottom Up*. Cambridge: MIT Press.

Estep, D. Q. and K. E. M. Bruce (1981). The concept of rape in non-humans: A critique. *Animal Behavior 29*, 1272–1273.

Ferguson, N., D. Cummings, S. Cauchemez, C. Fraser, S. Riley, A. Meeyai, S. Iamsirithaworn, and D. Burke (2005). Strategies for containing an emerging influenza pandemic in southeast asia. *Nature 437*, 209–214.

Fisher, R. (1930). *The Genetical Theory of Natural Selection*. Oxford, UK: Oxford University Press.

Fisher, R. (1957, 3 August). Letter. *British Medical Journal*, 297–8.

Fodor, J. A. (1983). *The Modularity of Mind: An Essay on Faculty Psychology*. Cambridge, MA: MIT Press.

Fodor, J. A. and J. Piattelli-Palmarini (2010). *What Darwin Got Wrong*. London: Profile Books.

Fogel, L. J., A. J. Owens, and M. J. Walsh (1966). *Artificial Intelligence through Simulated Evolution*. New York: Wiley.

Foot, P. (1978). *Virtues and Vices, and Other Essays in Moral Philosophy*. Oxford: Blackwell.

Fox, G. and C. Bruce (2001). Conditional fatherhood: Identity thoery and parental investment theory as alternative sources of explanation of fathering. *Journal of Marriage and Family 63*(2), 394–403.

Franklin, A. (1986). *The Neglect of Experiment*. Cambridge University Press.

Franklin, A. (1990). *Experiment, Right or Wrong*. New York: Cambridge University Press.

Fraser, N. and D. Kilgour (1986). Non-strict ordinal 2×2 games: A comprehensive computer-assisted analysis of the 726 possibilities. *Theory and Decision 20*(2), 99–121.

Freedman, D. and P. Humphreys (1999). Are there algorithms that discover causal structure? *Synthese 121*, 29–54.

Frigg, R. and J. Reiss (2009). The philosophy of simulation: hot new issues or same old stew? *Synthese 169*, 593–613.

Gaylord, R. and L. D'Andria (1998). *Simulating Society*. New York: Springer Verlag.

Gazzaniga, M. (1989). Organization of the human brain. *Science 245*, 947–952.

Geary, D. C. (1996). Biology, culture, and cross-national differences in mathematical ability. In R. J. Sternberg and T. Ben-Zeev (Eds.), *The nature of mathematical thinking*. Mahwah, NJ: L. Erlbaum Associates.

Giere, R. N. (1985). Philosophy of science naturalized. *Philosophy of Science 52*, 331–356.

Giere, R. N. (1999). Using models to represent reality. In L. Magnani, N. Nersessian, and P. Thagard (Eds.), *Model-Based Reasoning in Scientific Discovery*, pp. 41–57. New York: Kluwer.

Gigerenzer, G. and P. Todd (1999). *Simple Heuristics that Make Us Smart*. New York: Oxford University Press.

Gilovich, T., D. Griffin, and D. Kahneman (2002). *Heuristics and Biases: The Psychology of Intuitive Judgement.* Cambridge: Cambridge University.

Gilpin, M. E. (1975). *Group Selection in Predator-prey Communities.* Princeton University Press.

Gintis, H. (2000). Strong reciprocity and human sociality. *Journal of Theoretical Biology 206*, 169–179.

Glimcher, P. (2004). *Decisions, Uncertainty, and the Brain: The Science of Neuroeconomics.* The MIT Press.

Goldberg, D. E. (1989). *Genetic Algorithms in Search, Optimization, and Machine Learning.* Reading, MA: Addison-Wesley.

Goldsmith, T. (2004). Ageing as an evolved characteristic – Weismann's theory reconsidered. *Medical Hypotheses 62*, 304–308.

Goldspink, C. (2002). Methodological implications of complex systems approaches to sociality: Simulation as a foundation for knowledge. *Journal of Artificial Societies and Social Simulation 5*(1).

Goldstine, H. H. (1993). *The Computer from Pascal to von Neumann.* Princeton: Princeton University.

Gooding, D. C. and T. R. Addis (2008). Modelling experiments as mediating models. *Foundations of Science 13*, 17–35.

Goodman, N. (1956). *Fact, Fiction, and Forecast.* Indianapolis: Bobbs-Merrill.

Gould, S. (2002). *The Structure of Evolutionary Theory.* Harvard University Press.

Gould, S. J. and R. C. Lewontin (1979). The spandrels of San Marco and the Panglossian paradigm: A critique of the adaptationist programme. *Royal Society of London Proceedings Series B 205*, 581–598.

Gowaty, P. A. (1982). Sexual terms in sociobiology: Emotionally evocative and paradoxically, jargon. *Animal Behavior 30*, 630–631.

Gowaty, P. A. and N. Buschhaus (1998). Ultimate causation of aggressive and forced copulation in birds: Female resistance, the CODE hypothesis and social monogamy. *American Zoologist 38*, 207–225.

Grimm, V. (1999). Ten years of individual-based modelling in ecology: what have we learned and what could we learn in future? *Ecological Modelling 115*, 129–148.

Grimm, V. and S. Railsback (2005). *Individual-based Modelling and Ecology*. Princeton: Princeton University Press.

Grimm, V., E. Revilla, U. Berger, F. Jeltsch, W. M. Mooij, S. F. Railsback, H. Thulke, J. Weiner, T. Wiegand, and D. L. DeAngelis (2005). Pattern-oriented modeling of agent-based complex systems: Lessons from ecology. *Science 310*, 987–991.

Hamilton, W. (1964). The genetical evolution of social behavior I & II. *Journal of Theoretical Biology 7*, 1–16 & 17–52.

Hamilton, W. (1975). Innate social aptitudes of man: An approach from evolutionary genetics. In R. Fox (Ed.), *Biosocial Anthropology*, New York, pp. 133–155. John Wiley and Sons.

Hartmann, S. (1996). The world as a process: Simulation in the natural and social sciences. In R. Hegselmann, U. Müller, and K. Troitzsch (Eds.), *Modelling and Simulation in the Social Sciences from the Philosophy of Science Point of View*, pp. 77–100. Kluwer.

Hauser, M. (2006). *Moral Minds: How Nature Designed Our Universal Sense of Right and Wrong*. New York: Harper Collins.

Heath, K. M. and C. Hadley (1998). Dichotomous male reproductive strategies in a polygynous human society: Mating versus parental effort. *Current Anthropology 39*(3), 369–374.

Hegselmann, R. and U. Krause (2006). Truth and cognitive division of labour. *Journal of Artificial Societies and Social Simulation 9*. http://jasss.soc.surrey.ac.uk.

Hey, J. (2006). On the failure of modern species concepts. *Trends in Ecology and Evolution 21*, 447–450.

Hinton, G. E. and S. J. Nowlan (1987). How learning can guide evolution. *Complex Systems 1*, 495–502.

Holland, J. (1975). *Adaptation in Natural and Artificial Systems*. Ann Arbor: University of Michigan Press.

Howson, C. and P. Urbach (2006). *Scientific Reasoning: The Bayesian Approach*. Chicago: Open Court.

Hrdy, S. B. (1979). Infanticide among animals: A review, classification and examination of the implications for the reproductive strategies of females. *Ethology and Sociobiology 1*, 13–40.

Hrdy, S. B. (1999). *Mother Nature: A History of Mothers, Infants, and Natural Selection*. New York: Pantheon Books.

Hull, D. (1988). *Science as a Process*. Chicago: Chicago University Press.

Hume, D. (1739). *A Treatise of Human Nature*. London: Penguin (1984). Edited by E. C. Mossner.

Humphreys, P. (1991). Computer simulations. In *Philosophy of Science Association 1990*, Volume 2, pp. 497–506.

Humphreys, P. (1993). Numerical experimentation. In P. Humphreys (Ed.), *Patrick Suppes: Scientific Philosopher*, Volume 2. Dordrecht: Kluwer.

Humphreys, P. (2004). *Extending Ourselves: Computational Science, Empiricism, and Scientific Method*. Oxford: Oxford University Press.

Hutcheson, F. (1738). *An Inquiry into the Original of Our Ideas of Beauty and Virtue: In Two Treatises: I. Concerning Beauty, Order, Harmony, Design: II. Concerning Moral Good and Evil*. (4th ed.). Farnsborough: Greg International Publishers (1969).

Huth, C. and O. Pellmyr (2000). Pollen-mediated selective abortion in yuccas and its consequences for the plant-pollinator mutualism. *Ecology 81*(4), 1100–1107.

Huxley, J. S. (1927). *Religion without Revelation*. London: Ernest Benn.

Huxley, J. S. (1942). *Evolution, the Modern Synthesis*. London: Allen and Unwin.

Huxley, T. and J. Huxley (1947). *Evolution and Ethics: 1893-1943*. London: Pilot.

Jablonka, E. and E. Szathmáry (1995). The evolution of information storage and heredity. *Trends in Ecology and Evolution 10*, 206–211.

Jaffe, K. (2002). An economic analysis of altruism: who benefits from altruistic acts? *Journal of Artificial Societies and Social Simulation* 5(3).

James, W. (1890). *Principles of Psychology*. New York: Henry Holt.

James, W. (1891). The moral philosopher and the moral life. *International Journal of Ethics 1*, 330–354.

Janus, L. (1997). *The Enduring Effects of Prenatal Experience: Echoes from the Womb*. Northvale, NJ: Jason Aronson.

Joiner Jr, T. E., J. W. Pettit, R. L. Walker, Z. R. Voelz, J. Cruz, M. D. Rudd, and D. Lester (2002). Perceived burdensomeness and suicidality: Two studies on the suicide notes of those attempting and those completing suicide. *Journal of Social and Clinical Psychology 21*, 531–545.

Kahneman, D., P. Slovic, and A. Tversky (1982). *Judgment Under Uncertainty: Heuristics and Biases*. Cambridge: Cambridge University.

Kant, I. (1909). Fundamental principles of the metaphysic of morals. In T. K. Abbott (Ed.), *Kants critique of practical reason and other works on the theory of ethics*. London: Longmans.

Keller, E. F. (2003). *Making Sense of Life: Explaining Biological Development with Models, Metaphors, and Machines*. Cambridge, MA: Harvard University Press.

Kim, J. (1993). *Supervenience and Mind: Selected Philosophical Essays*. Cambridge: Cambridge University Press.

Kimmel, M. (2003). An unnatural history of rape. In C. B. Travis (Ed.), *Evolution, Gender, and Rape*, pp. 221–233. Cambridge Mass: MIT Press.

Kitcher, P. (1985). *Vaulting Ambition: Sociobiology and the Quest for Human Nature*. Cambridge, MA: MIT Press.

Knobe, J. and J. Doris (2008). Strawsonian variations: Folk morality and the search for a unified theory. In J. D. et al. (Ed.), *Handbook of Moral Psychology*. Oxford: Oxford University Press.

Korb, K. B. (1994). Stephen J. Gould on intelligence. *Cognition 52*, 111–123.

Korb, K. B. and S. Mascaro (2009). The philosophy of computer simulation. In C. Glymour, W. Wei, and D. Westerstahl (Eds.), *Logic, Methodology and Philosophy of Science: Proceedings of the Thirteenth International Congress*, pp. 306–325. College Publications.

Korb, K. B. and A. E. Nicholson (2010). *Bayesian Artificial Intelligence* (2nd ed.). Boca Raton, FL: CRC/Chapman & Hall.

Korb, K. B. and E. Nyberg (2006). The power of intervention. *Minds and Machines 16*, 289–302.

Kueppers, G., J. Lenhard, and T. Shinn (2006). Computer simulation: Practice, epistemology, and social dynamics. In J. Lenhard, G. Kueppers, and T. Shinn (Eds.), *Simulation: Pragmatic Construction of Reality*. Dordrecht: Springer.

Langton, C. (Ed.) (1989). *Artificial Life: The Proceedings of an Interdisciplinary Workshop on the Synthesis and Simulation of Living Systems*, Redwood City, CA. Addision-Wesley.

Leroi, A. M., A. K. Chippindale, and M. R. Rose (1994). Long-term laboratory evolution of a genetic life-history trade-off in *Drosophila Melanogaster*. *Evolution 48*, 1244–1257.

Levin, B. R. and W. L. Kilmer (1974). Interdemic selection and the evolution of altruism: A computer simulation study. *Evolution 28*(4), 527–545.

Lewis, D. (1986). *Philosophical Papers*, Volume II. Oxford: Oxford University Press.

Lewontin, R. (2000). *It Ain't Necessarily So: The Dream of the Human Genome and Other Illusions*. London: Granta.

Longini, I., M. Halloran, A. Nizam, and Y. Yang (2004). Containing pandemic influenza with antiviral agents. *American Journal of Epidemiology 159*, 623.

Lycett, J. E. and R. Dunbar (1999). Abortion rates reflect optimisation of parental investment strategies. *Proceedings of the Royal Society, B.: Biological Sciences 266*(1436), 2355–2358.

Machery, E., R. Mallon, S. Nichols, and S. P. Stich (2004). Semantics, cross-cultural style. *Cognition 92*, 1–12.

MacIntyre, A. (1981). *After Virtue: A Study in Moral Theory*. London: Duckworth.

Mascaro, S., K. B. Korb, and A. E. Nicholson (2001). Suicide as an evolutionary stable strategy. In *Advances in Artificial Life, 6th European Conference*, Prague, Czech Republic, pp. 358–361.

Mascaro, S., K. B. Korb, and A. E. Nicholson (2005). An ALife investigation on the origins of dimorphic parental investments. In H. A. Abbass, T. Bossamaier, and J. Wiles (Eds.), *Proceedings of the Australian Conference on Artificial Life*, pp. 171–185.

Maynard Smith, J. (1976). Group selection. *Quarterly Review of Biology 51*, 277–283.

Mayr, E. (1963). *Animal Species and Evolution*. Cambridge: Harvard University Press.

Mayr, E. (1976). *Evolution and the Diversity of Life*. Cambridge, MA: Harvard University.

McBride, W. Z. (1991). Spontaneous abortion. *American Family Physician 43*, 175–182.

McCormack, J. (2005). A developmental model for generative media. In M. S. Capcarrere, A. A. Freitas, P. J. Bentley, C. G. Johnson, and J. Timmis (Eds.), *Proceedings of the 8th European Conf. on Advances in Artificial Life*, pp. 88–97.

McKenna, M. (2009). Compatibilism. http://plato.stanford.edu/ (Accessed: 15 July, 2010).

Meadows, D. (1972). *The Limits of Growth: A Report for the Club of Rome*. New York: Earth Island.

Medawar, P. (1952). *An Unsolved Problem in Biology*. London: H.K. Lewis.

Mitchell, M. and S. Forrest (1994). Genetic algorithms and artificial life. *Artificial Life 1*(3), 267–289.

Mitteldorf, J. (2002). Multilevel selection and the evolution of predatory restraint. In R. K. Standish, M. A. Bedau, and H. A. Abbass (Eds.), *Artificial Life VIII: Proceedings of the Eighth International Conference on the Simulation and Synthesis of Living Systems*, pp. 146–152.

Mitteldorf, J. (2004). Aging selected for its own sake. *Evolutionary Ecology Research 7*, 1–17.

Mitteldorf, J. (2006). Chaotic population dynamics and the evolution of aging. *Evolutionary Ecology Research 3*, 561–574.

Moore, G. E. (1903). *Principia Ethica*. Cambridge: Cambridge University Press.

Morgan, M. (2002). Model experiments and models in experiments. In L. Magnani and N. J. Nersessian (Eds.), *Model Based Reasoning: Science, Technology, Values*, pp. 41–58. Springer.

Nichols, S. and J. M. Knobe (Eds.) (2008). *Experimental Philosophy*. Oxford: Oxford University.

Norton, S. D. and F. Suppe (2001). Why atmospheric modelling is good science. In C. Miller and P. Edwards (Eds.), *Changing the Atmosphere: Expert Knowledge and Environmental Governance*, pp. 67–105. Cambridge, MA: MIT Press.

Nowak, M. and R. May (1992). Evolutionary games and spatial chaos. *Nature 359*(6398), 826–829.

Nowak, M. and K. Sigmund (1993). A strategy of win-stay, lose-shift that outperforms tit-for-tat in the prisoner's dilemma game. *Nature 364*(6432), 56–58.

Nozick, R. (1974). *Anarchy, State and Utopia*. Oxford: Blackwell.

Odling-Smee, F. J., N. Laland, and M. W. Feldman (2003). *Niche Construction*. Princeton: Princeton University.

Ofria, C. and C. O. Wilke (2004). Avida: A software platform for research in computational evolutionary biology. *Journal of Artificial Life 10*, 191–229.

Oreskes, N., K. Shrader-Frechette, and K. Belitz (1994, February). Verification, validation and confirmation of numerical models in the earth sciences. *Science 263*(5147), 641–646.

Over, D. E. (Ed.) (2003). *Evolution and the Psychology of Thinking: The Debate*. New York: Psychology Press.

Parfit, D. (1984). *Reasons and Persons.* Oxford: Clarendon Press.

Pargellis, A. (1996). The evolution of self-replicating computer organisms. *Physica D 98*(1), 111–127.

Pellmyr, O. and C. J. Huth (1994). Evolutionary stability of mutualism between yuccas and yucca moths. *Nature 372*(6503), 257–260.

Penrose, R. (1999). *The Emperor's New Mind: Concerning Computers, Minds, and the Laws of Physics* (2nd ed.). Oxford: Oxford University.

Pinker, S. (1994). *The Language Instinct.* New York: W. Morrow and Co.

Pinker, S. (2002). *The Blank Slate: The Modern Denial of Human Nature.* New York: Viking.

Popper, K. (1953). *The Open Society and Its Enemies*, Volume I. Routledge and Kegan Paul.

Price, G. R. (1970). Selection and covariance. *Nature 227*, 520–521.

Pritsker, A. A. B. (1979). Compilation of definitions of simulation. *Simulation 33*, 61–63.

Pritsker, A. A. B. (1984). *Introduction to Simulation and SLAM II.* New York: John Wiley & Sons.

Putnam, H. (1975). *Mind, Language and Reality: Philosophical Papers*, Volume 2. Cambridge, MA: MIT Press.

Quine, W. and J. Ullian (1978). *The Web of Belief.* New York: Random House.

Quine, W. V. O. (1969). Epistemology naturalized. In *Ontological Relativity and Other Essays.* Columbia University.

Racynski, S. and A. Bargiela (2007). *Modeling and Simulation: Computer Science of Illusion.* Research Studies Press. http://portal.acm.org/.

Railsback, S. F., R. H. Lambersion, B. C. Harvey, and W. E. Duffy (1999). Movement rules for individual-based models of stream fish. *Ecological Modelling 123*, 73–89.

Ramsey, F. P. (1931). Truth and probability. In R. B. Braithwaite (Ed.), *The Foundations of Mathematics and Other Logical Essays*. London: K. Paul, Trench, Trubner & co.

Randall, D. A., R. A. Wood, S. Bony, R. Colman, T. Fichefet, J. Fyfe, V. Kattsov, A. Pitman, J. Shukla, J. Srinivasan, R. J. Stouffer, A. Sumi, and K. E. Taylor (2007). Climate models and their evaluation. In S. Solomon, D. Qin, M. Manning, Z. Chen, M. Marquis, K. B. Averyt, M. Tignor, and H. L. Miller (Eds.), *Climate Change 2007: The Physical Science Basis. Contribution of Working Group I to the Fourth Assessment Report of the Intergovernmental Panel on Climate Change*. Cambridge: Cambridge University Press.

Rawls, J. (1972). *A Theory of Justice*. Oxford: Oxford University Press.

Ray, T. S. (1991). An approach to the synthesis of life. In C. Langton, C. Taylor, J. D. Farmer, and S. Rasmussen (Eds.), *Artificial Life II, Santa Fe Institute Studies in the Sciences of Complexity*, pp. 371–408. Redwood City, CA: Addison-Wesley.

Rechenberg, I. (1971). *Evolutionstrategie: Optimierung technischer Systeme nach Prinzipien der biologischen Evolution*. Ph. D. thesis, Technische Universität München.

Reddy, R. (1987). Epistemology of knowledge-based systems. *Simulation 48*, 161–170.

Reynolds, C. W. (1987). Flocks, herds, and schools: A distributed behavioral model. *Computer Graphics 21*, 15–34.

Ridley, M. (1993). *The Red Queen: Sex and the Evolution of Human Nature*. London: Viking.

Roberts, C. and C. Lowe (1975). Where have all the conceptions gone? *Lancet 1*, 498–99.

Robson, A. and L. Samuelson (2008). The evolutionary foundations of preferences. In J. Benhabib, A. Bisin, and M. Jackson (Eds.), *The Social Economics Handbook*. Elsevier. forthcoming.

Rohrlich, F. (1991). Computer simulation in the physical sciences. In *PSA 1990*, Volume II, pp. 507–518.

Russell, S. and P. Norvig (2010). *Artificial Intelligence: A Modern Approach* (3rd ed.). Englewood Cliffs, NJ: Prentice-Hall.

Sample, I. (2010, 20 May). Craig Venter creates synthetic life form. Guardian. http://www.guardian.co.uk/science (Accessed: 27 July 2010).

Samuelson, L. and J. Swinkels (2006). Information, evolution and utility. *Theoretical Economics 1*, 119–142.

Scheffler, S. (1982). *The Rejection of Consequentialism*. Oxford: Clarendon Press.

Schwefel, H.-P. (1981). *Numerical Optimization of Computer Models*. New York: Wiley.

Scogings, C. and K. Hawick (2008). Altruism amongst spatial predator-prey animats. In S. Bullock, J. Noble, R. Watson, and M. A. Bedau (Eds.), *Artificial Life XI: Proceedings of the Eleventh International Conference on the Simulation and Synthesis of Living Systems*, Volume 11, Cambridge, pp. 537–544. MIT Press.

Segerstrale, U. (2000). *Defenders of the Truth: The Battle for Science in the Sociobiology Debate and Beyond*. Oxford: Oxford University.

Selfridge, O. G. (1959). Pandemonium: A paradigm for learning. In *Proceedings of the Symposium on the Mechanization of Thought Processes*. London: Her Majesty's Stationery Office.

Sidgwick, H. (1907). *The Methods of Ethics*. London: Macmillan.

Singer, P. (1976). *Animal Liberation*. London: Jonathan Cape.

Singer, P. (1993). *Practical Ethics*. Cambridge: Cambridge University Press.

Singer, P. (1994). Introduction to 'Ethics'. In P. Singer (Ed.), *Ethics*. Oxford: Oxford University Press.

Skulachev, V. (1997). Ageing is a specific biological function rather than the result of a disorder in complex living systems: Biochemical evidence in support of Weismann's hypothesis. *Biochemistry (Moscow) 62*, 1191–1195.

Skyrms, B. (2004). *The Stag Hunt and the Evolution of Social Structure.* Cambridge: Cambridge Univ.

Skyrms, B. and R. Pemantle (2000). A dynamic model of social network formation. In *Proceedings of the National Academy of Sciences*, Volume 97, pp. 9340–9346. National Academy of Sciences.

Sober, E. and D. S. Wilson (1998). *Unto Others.* Cambridge: Havard University Press.

Sociobiology Study Group of Science for the People (1978). Sociobiology: Another biological determinism. In A. L. Caplan (Ed.), *The Sociobiology Debate*, pp. 280–290. New York: Harper & Row.

Spencer, H. (1864). *Principles of Biology.* D. Appleton.

Sterelny, K. and P. E. Griffiths (1999). *Sex and Death: An Introduction to the Philosophy of Biology.* Chicago: University of Chicago.

Stotz, K. and P. E. Griffiths (2004). Genes: Philosophical analyses put to the test. *History and Philosophy of the Life Sciences 26*, 5–28.

Stroud, P., S. Del Valle, S. Sydoriak, J. Riese, and S. Mniszewski (2007). Spatial dynamics of pandemic influenza in a massive artificial society. *Journal of Artificial Societies and Social Simulation 10.* http://jasss.soc.surrey.ac.uk.

Sumner, L. (1981). *Abortion and Moral Theory.* Princeton, NJ: Princeton University Press.

Suppe, F. (1977). *The Structure of Scientific Theories.* Urbana: University of Illinois Press.

Symons, D. (1979). *The Evolution of Human Sexuality.* New York: Oxford University Press.

Tennant, N. (1999). Sex and the evolution of fair-dealing. *Philosophy of Science 66*(3), 391–414.

Thornhill, R. and C. T. Palmer (2000). *A Natural History of Rape.* London: MIT Press.

Thornhill, R. and N. Thornhill (1983). Human rape: An evolutionary analysis. *Ethology and Sociobiology 4*, 137–173.

Tooby, J. and L. Cosmides (1992). The psychological foundations of culture. In J. H. Barkow, L. Cosmides, and J. Tooby (Eds.), *The Adapted Mind: Evolutionary Psychology and the Generation of Culture*. Oxford University Press, New York.

Toyama, M. (2001). Adaptive advantages of matriphagy in the foliage spider, *Chiracanthium Japonicum*. *Journal of Ethology 19*, 69–74.

Trivers, R. (1971). The evolution of reciprocal altruism. *Quarterly Review of Biology 46*, 35–57.

Trivers, R. (1972). Parental investment and sexual selection. In B. Campbell (Ed.), *Sexual Selection and the Descent of Man*, pp. 136–179. London: Heinemann.

Trivers, R. (1974). Parent-offspring conflict. *American Zoologist 11*, 249–264.

Trivers, R. and D. Willard (1973). Natural selection of parental ability to vary the sex ratio of offspring. *Science 179*(4068), 90–92.

Verrell, P. A. and N. R. McCabe (1990). Major histocompatibility antigens and spontaneous abortion: an evolutionary perspective. *Medical Hypotheses 32*(3), 235–238.

Volterra, V. (1931). Variations and fluctuations of the number of individuals in animal species living together. In R. Chapman (Ed.), *Animal Ecology*.

von Neumann, J. (1951). The general and logical theory of automata. In A. H. Taub (Ed.), *John von Neumann: Collected works.*, pp. 288–328. New York: Pergamon Press.

von Neumann, J. and O. Morgenstern (1944). *Theory of Games and Economic Behavior*. Princeton, NJ: Princeton University Press.

Ward, P. (1995). *The End of Evolution: Dinosaurs, Mass Extinction and Biodiversity*. London: Weidenfeld and Nicholson.

Wason, P. C. (1966). Reasoning. In B. M. Foss (Ed.), *New Horizons in Psychology (Vol. 1)*. Harmondsworth: Penguin.

Weindruch, R. and R. Walford (1986). *The Retardation of Aging and Disease by Dietary Restriction*. Springfield, Illinois: Thomas.

Weismann, A. (1889). *Essays upon Heredity and Kindred Biological Problems*. Oxford: Clarendon Press.

Wiener, N. (1948). *Cybernetics: Or, Control and Communication in the Animal and Machine*. Cambridge, MA: MIT Press.

Wikipedia (2010). Utilitarianism. http://en.wikipedia.org/wiki/Utilitarianism (Accessed: 6 July, 2010).

Williams, G. (1957). Pleiotropy, natural selection, and the evolution of senescence. *Evolution 11*, 398–411.

Williams, G. (1975). *Sex and Evolution*. Princeton, NJ: Princeton University Press.

Williams, G. C. (1966). *Adaptation and Natural Selection*. Princeton, NJ: Princeton University Press.

Wilson, D. S. (1980). *The Natural Selection of Populations and Communities*. Menlo Park: Benjamin/Cummings.

Wilson, D. S. (1997). Introduction: Multilevel selection theory comes of age. *American Naturalist 150, Supplement*, S1–S4. This introduces a special issue on multilevel selection.

Wilson, D. S. (2005). Human groups as adaptive units: toward a permanent consensus. In P. Carruthers, S. Laurence, and S. Stich (Eds.), *The Innate Mind: Volume 2, Culture and Cognition*, pp. 78–90. Oxford University Press.

Wilson, E. O. (1975). *Sociobiology: The New Synthesis*. Cambridge, MA: Harvard University Press.

Wilson, E. O. (1977). Biology and the social sciences. *Daedalus 106*, 127–140.

Wilson, E. O. (1978a). *On Human Nature*. Cambridge: Harvard University.

Wilson, E. O. (1978b). Introduction: What is sociobiology? In M. S. Gregory, A. Silvers, and D. Sutch (Eds.), *Sociobiology and Human Nature: An Interdisciplinary Critique and Defense*, pp. 1–12. San Francisco: Jossey-Bass.

Wilson, S. W. (1989). The genetic algorithm and simulated evolution. In C. Langton (Ed.), *Artificial Life: Proceedings of an Interdisciplinary Workshop on the Synthesis and Simulation of Living Systems*, Redwood City, CA. Addison-Wesley.

Winsberg, E. (1999). Sanctioning models: The epistemology of simulation. *Science in Context 12(2)*, 275–292.

Winsberg, E. (2001). Simulations, models, and theories: Complex physical systems and their representations. In *Proceedings of the 2000 Biennial Meetings of the Philosophy of Science Association (Supplement to Philosophy of Science Vol. 68 No. 3)*, pp. S442–S454.

Winsberg, E. (2003). Simulated experiments: methodology for a virtual world. *Philosophy of Science 70(1)*, 105–125.

Wolfram, S. (2002). *A New Kind of Science*. Wolfram Media.

Wright, S. (1922). Coefficients of inbreeding and relationship. *American Naturalist 56*, 330–338.

Wynne-Edwards, V. (1962). *Animal Dispersion in Relation to Social Behaviour*. Edinburgh: Oliver and Boyd.

Yaeger, L., V. Griffith, and O. Sporns (2008). Passive and driven trends in the evolution of complexity. In *Artificial Life XI: Proceedings of the Tenth International Conference on the Simulation and Synthesis of Living Systems*, pp. 725–732.

Yaeger, L. and O. Sporns (2006). Evolution of neural structure and complexity in a computational ecology. In *Artificial Life X: Proceedings of the Tenth International Conference on the Simulation and Synthesis of Living Systems*, pp. 330–336.

Zeigler, B. (1976). *Theory of Modeling and Simulation*. New York: Wiley-Interscience.

Glossary

Adaptation Adaptation either refers to an evolutionary process or to a trait resulting from such a process. The adaptive process fixes traits in a population through the action of natural selection. Adaptive traits are structures, behaviors or strategies that increase fitness and so are likely to be fixed in the population by natural selection.

Agent There are two senses of 'agent' employed in the literature that we refer to. (1) An agent is a behaving system with intentionality — i.e., with beliefs, desires and purposes — and so capable of moral responsibility. (2) An agent is a behaving system — i.e., a system that has some dynamics associated with it. Pieces of code interpreted in **computer processes** which have little or no cognitive ability are often called "agents" by their designers, in much the same hopeful way they call their tiny representational abilities "knowledge".

Agent-Based Modeling (ABM) The study of social systems using computer simulation of individuals within an environment; a form of **ALife** simulation.

Aging Aging is the general deterioration of an organism via internal causes, leading to its eventual death.

Algorithm An algorithm is a type of procedure for implementing a function from a range of possible inputs to particular outputs. The three defining features of an algorithm are: (1) definiteness (its steps must be "primitive" and well understood, as in the steps of a Turing machine); (2) finiteness (it must stop); (3) functionality (its output must always be the same given the same input). An algorithm may be implemented by any number of distinct programs (Turing machines).

Stochastic algorithms incorporate the appearance of indeterminism, by using pseudo-random number generators to sample from probability distributions. This introduces variation into some aspect of the performance of the program, for example in mutations during artificial reproduction. Nevertheless, these are algorithms, since the pseudo-random number generators are deterministic.

Allele An alternative form of a DNA sequence at some **locus** in the chromosome.

Allopatric Speciation Speciation that occurs after a population has divided into isolated, geographical groups.

Altruism An altruistic act is one which harms the acting agent while benefitting others. The actor incurs negative utility while others derive positive utility. See also **Biological Altruism**.

Artificial Life (ALife) The study of the basic processes of life, real or possible, using computer simulation or other technology.

Artificial Neural Network (ANN) An artificial neural network is a model of neural processing in the brain. An ANN is composed of interconnected neurons, where each neuron is a function taking inputs either externally or from other neurons, and yielding an output. Commonly, the function involves a weighted sum of the inputs which is then passed through a continuous equivalent of a threshold function such as the logistic function (called the neuron's activation function) to yield the output.

Bayesianism, Bayesian A Bayesian takes probability functions to be the (or a) key means to represent and reason about uncertainty. Bayesians use probability to represent subjective degrees of belief in propositions and generally advocate conditionalization — adopting the conditional probability function based upon the evidence acquired — as a useful method of modeling evidential learning. This approach can be applied to any domain in which probabilities (or degrees of belief) arise, such as in the expected value calculations of decision theory, scientific theory **confirmation** and causal modeling.

Bayesian Network (BN) A Bayesian network is a graphical means of representing probability distributions and associated software for performing evidential updating. The graphs are sets of variables (or nodes) connected by directed arcs without cycles (i.e., directed acyclic graphs).

GLOSSARY

A directed arc represents a conditional probabilistic dependency between a parent variable (the arc source) and the child variable (the arc destination), such that the state of a child depends on the state of all its parents.

Biological Altruism An **allele** exhibits biological altruism if and only if its presence reduces the individual fitness of its owner while increasing the individual fitnesses of others. Phenotypic traits (such as behaviors) may analogously be considered altruistic if they do the same. See **Altruism**.

By-product (Piggyback Trait) A by-product is a trait that is evolutionarily neutral or harmful but has survived the evolutionary process because of its necessary connection with other, adaptative, traits. See **Adaptation**.

Cellular Automaton A cellular automaton is a grid of cells (of any finite number of dimensions, but frequently two dimensions) in which each cell can take on a finite number of states (often just the two states "on" and "off"). The state of each cell changes over (discrete) time and is a function of its own state and that of neighbouring cells from the previous time step. The most famous cellular automaton is The Game of Life by John Conway.

Co-evolution Co-evolution describes scenarios in which two evolving species inhabit each others' environments and affect each others' evolutionary histories. Commonly discussed co-evolutionary scenarios include predators and prey, hosts and parasites, and flowers and pollinators.

Computer Program A program is a sequence of instructions for a (virtual) machine. The beginning and end of a program are its first and last sentences.

Computer Process A computer process is a process run by a (virtual) machine which interprets a **computer program**, generating outputs given some inputs. Like all processes, a computer process has a temporal beginning and end; these are usually not directly related to the beginning and end of the program being interpreted.

Confirmation, Bayesian Confirmation Theory In the philosophy of science confirmation refers to cases in which evidence is discovered that increases the probability of a given hypothesis or theory. The opposite is disconfirmation, in which the probability of the hypothesis or theory is decreased by the new evidence.

Consequentialism Consequentialism holds that the effects or consequences of an act must be taken into account when determining the ethical value of an act. See **Deontological Ethics**.

Cooperation Cooperation is coordinated activity aimed at achieving a common goal. The common goal often provides mutual benefits, but this is not required. Since participants must be capable of possessing goals, they must also possess agency. See **Agent** (1).

Cultural Evolution In the context of evolutionary biology, cultural evolution is the change in culture over time due to processes analogous with those found in biological evolution: namely, **heritability**, selection and variation. Cultural evolution substitutes elements of culture, such as ideas, theories and customs, learning and mistakes and creativity, for the corresponding elements of biological evolution (such as genes, reproduction and mutation). In most such models, the distinction between genotype and phenotype is either absent or ill-defined.

Decision Theory Decision theory examines how choices are or can be made under uncertainty. At the heart of decision theory is the idea of the expected value of a choice (a sum of all the possible outcomes, good, bad or neutral, weighted by the probability of each outcome), allowing choices to be placed on an interval scale. The value of outcomes is typically assessed in accordance with utility theory. See **Utility Theory**.

Decision Tree (Classification Tree) A decision tree is a classifier function that operates by recursively classifying the input into increasingly finer-grained classes. It is represented by a tree (typically visualized upside-down), with branches, leaves (or output) and a single root. Branches split the input data into two or more classes; once a branch identifies which class the input falls into, it passes execution to either the next branch for that class or, as a stopping condition, the leaf associated with that class. See **Production Rule**.

Deme A deme is a locally isolated and interbreeding sub-group of a population.

GLOSSARY

Deontological Ethics Deontological ethics holds that goodness inheres in either acts, duties, rules or rights. See **Consequentiaism**.

Descriptive Ethics The study of what people believe about how we ought to behave. See **Normative Ethics**.

Dominant Strategy A strategy (or action) dominates another in game theory if and only if the **utility** of each possible state on the first strategy is greater than that for the same state on the second.

Egoism Ethical egoism (or just 'egoism') is the belief that people ought to do what is in their own self-interest. See **Hedonism**.

Emergence, Emergent Property Emergent properties of a system A are its higher-level properties that cannot be defined in terms of the **supervenience** base B (and its properties) which implement it. See **Supervenience**.

Environment of Evolutionary Adaptation (EEA) The circumstances in which a characteristic, or set of characteristics, evolved. In particular, this is used to refer to the prehistoric circumstances in which the human mind evolved, contrasting them with the historic circumstances in which the human mind is currently operating.

Evolutionary Algorithm An evolutionary algorithm uses the operators of evolution (selection, reproduction and mutation) to successively modify or (in optimization problems) improve the current population of entities or solutions. Examples of evolutionary algorithms include evolution strategies, evolutionary programming, genetic programming and genetic algorithms. The main difference between these algorithms is what representations are used in the genomes or members of the population — whether bit strings, numbers, graphs, vectors, state machines or programs. See **Genetic Algorithm**.

Evolutionary Ethics Evolutionary ethics is the attempt to found ethical norms on evolutionary history.

Evolutionary Psychology Evolutionary psychology is the study of animal behavior from an evolutionary perspective. More particularly, it is a school of thought in such studies characterized by the beliefs that most behavior has an evolutionary explanation, that behaviors have arisen in "evolutionary environments of adaptation" (EEA), and that cognitive functions have often evolved in semi-independent modules. See **Sociobiology**.

Evolutionarily Stable Strategy (ESS) An evolutionarily stable strategy is a behavior determined (partially or fully) by genetics such that, if adopted by all members of a population, no alternative strategy can invade and replace the ESS. In other words, under the circumstances (where most of the population has adopted the ESS), the ESS is fitter than its alternatives and so resists invasion. The behaviors may be pure or **mixed strategies**.

Filial Infanticide The killing of one's own offspring shortly after birth, while the offspring is still a dependant.

Fisher's Reproductive Value The reproductive value of an organism of a given age reflects the expected future number of offspring of the organism. Reproductive value is calculated by summing (from the organism's current age onwards) the probability of reaching a given age multiplied by the average number of offspring produced by an individual at that age and dividing the sum by the average population fitness.

Fitness Individual fitness is often defined as the expected number of offspring reaching maturity. We prefer to define it as the expected number of descendants, which accommodates both uncertainty about the future (through probabilistic expectation) and issues about the viability and fertility of offspring. For practical measurements, we often substitute descendants over two generations. See **Inclusive Fitness**.

Gene A sequence of DNA which codes for some protein(s).

Genetic Algorithm (GA) A genetic algorithm is a type of optimizing evolutionary algorithm that operates on a population of chromosomes (traditionally, raw bit strings) without any environment. The GA handles the reproduction, mutation and selection of these genomes. Selection for reproduction is directed by an artificial fitness function ("objective function") and occurs between discrete, successive generations. Evolutionary ALife simulations, by contrast, embed genomes within interacting agents that exist in a wider environment, where fitness is determined (as in biology) by agents' abilities to survive and reproduce.

GLOSSARY

Group Selection Group selection is selective pressure arising from the differential ability of groups to propagate themselves by establishing new groups, whether from greater rates of founding colonies, greater group longevity, or both. Given a correlation between this kind of group fitness with allele frequencies across groups, allele representations within the total population will tend to correspond to group selection pressure.

Hamming Distance Given two bit strings of equal length, the Hamming distance is the number of locations at which the two strings differ.

Hedonism Hedonism is the belief that pleasure is the only intrinsic good. Ethical hedonism (often just 'hedonism') merges egoism with hedonism yielding the belief that people ought to do what maximizes their own pleasure and minimizes their own pain. See **Egoism**.

Heritability Given a probability distribution over environments, the heritability of a trait is the amount of its variance which is explained genetically. (For standardized variables, this will be equal to one minus the amount of its variance explained by environmental variation.)

Homomorphism A homomorphism from system A to system B is a mapping of objects, functions and relations from system A onto system B such that all relations (and functions) within system A are preserved under the mapping in system B.

Inclusive Fitness Inclusive fitness measures the total fitness effects of an allele over a population. See **Kin Selection**.

Individual-Based Modeling (IBM) The study of biological systems using computer simulation of individuals within an environment; a form of **ALife** simulation.

Individual Selection Individual selection is differential natural selection pressure operating upon individual organisms through their different individual fitnesses. See **Kin Selection** and **Group Selection**.

Induced Abortion Induced abortion is the termination of a pregnancy by choice. See **Spontaneous Abortion**.

Isomorphism An isomorphism between system A and system B is a homomorphic mapping from system A to system B such that its inverse is a homomorphic mapping from system B to system A. See **Homomorphism**.

Kin Selection Kin selection identifies the impact of a phenotypic trait upon the fitness of the kin of the trait's bearer as the relevant factor for determining the spread of alleles coding for that trait. See **Inclusive Fitness**.

Levels of Selection Levels of selection refers to the type of selection pressure that may be active in an evolutionary system. The levels generally recognized as subject to significant selection are the gene, the cell, the individual, the group (**deme**) and the **species**. See **Multilevel Selection**.

Locus A location within DNA that repeatedly plays host to the same gene (and, so, some set of alleles).

Lotka-Volterra Model The Lotka-Volterra model is a simple model of predator-prey interactions described by a pair of differential equations. Each differential equation describes the rate of change in the numbers of either predators or prey given the current numbers of predators and prey, growth rates, encounters between predators and prey and death rates.

Mental Module A mental module is a substructure within the mind that has an evolved function (such as a module for language or facial recognition). In contrast to phrenology, mental modules need not (and are not generally expected) to correspond one to one with a physical substructure in the mind. See **Evolutionary Psychology**.

Metaethics The attempt to provide a theory of ethical study that allows us to choose between ethical systems.

Multilevel Selection Multilevel selection is selection pressure that acts simultaneously at multiple levels or, equivalently, on multiple biological units. The units generally recognized as subject to significant selection are the gene, the cell, the individual, the group (**deme**) and the **species**. It can be contrasted with the traditional (modern synthesis) view, in which genes are considered the only unit of consequence. See **Levels of Selection**.

Multiple Realizability See **Supervenience**.

GLOSSARY

Mixed Strategy A set of behaviors alternative to each other which are selected according to some probability distribution.

Naive Bayes A naive Bayes model is a simple Bayesian network used for (probabilistic) classification in which a single class node is the lone parent to all other nodes. The child nodes are independent of each other, given the class, and represent the attributes of some entity. See **Bayesian network**.

Nash Equilibrium A Nash equilibrium in game theory is a set of strategies such that no player can increase its utility by switching to an alternative strategy when the other players do not change strategy. See **Dominant Strategy**.

Naturalistic Fallacy Moore defined the naturalistic fallacy as the error of inferring an object's goodness from its natural properties. The term now generally refers to any attempt to directly derive 'ought' from 'is'.

Normative Ethics (Also, just "Ethics".) The study of how we ought to behave. See **Descriptive Ethics**.

Parapatric Speciation Speciation that occurs after a small group partially splinters from a larger group into a new, adjacent but not isolated geographical niche. See **Peripatric Speciation**.

Parental Investment Parental investment is any investment in an offspring that boosts that offspring's chance of survival but comes at the cost of investing in other offspring. It may refer to either material or behavioral support. See **Reproductive Strategy**.

Peripatric Speciation Speciation that occurs after a small group splinters from a larger group into an isolated geographical niche. See **Parapatric Speciation**.

Phyletic Gradualism Phyletic gradualism refers to evolutionary histories in which all evolutionary change occurs gradually (with no sudden jumps in phenotype space), including evolutionary change that gives rise to new species.

Pleiotropy Pleiotropy describes the genetic effect in which a single gene gives rise to multiple phenotypic traits. See **Polygenetic**.

Polygenetic A phenotypic trait is polygenetic if it is caused or influenced by multiple genes. See **Pleiotropy**.

Pop Sociobiology Pop Sociobiology is a term used by critics to describe sociobiology that relies heavily on just-so stories to explain the origins of modern human behavior. Fierce criticism of pop sociobiology lead to development of the methodologically more rigorous field of evolutionary psychology. See **Sociobiology** and **Evolutionary Psychology**.

Pre-adaptation A pre-adaptation is a trait that has evolved as an adaptation to one situation, but is subsequently put to a different use in an evolutionarily novel situation. See **Adaptation**.

Price Equation The Price equation is a generalization of kin and group selection models (and of selection models in general) which allows us to separate the selection effects acting within groups from those acting between groups. The equation is:

$$\bar{w}\Delta\bar{z} = \text{Cov}(w_i, z_i) + \text{E}(w_i \Delta z_i) \qquad (7.1)$$

On the right hand side, z is the character of interest (e.g., height, eye color or altruistic disposition) assumed to be representable by a real number, i identifies a subgroup of the population that shares the same value for z, z_i is the shared value itself (e.g., tall, blue or selfish), Δz_i is the change in this character from generation to generation, and w_i is the average absolute fitness of the subgroup i with trait z_i. On the left hand side, $\bar{w}$ is the average absolute fitness across the population overall and $\Delta \bar{z}$ is the average change from generation to generation in the character z over the population overall. By dividing through by $\bar{w}$, evolutionary change can be explicitly phrased in terms of relative fitness (i.e. $w_i/\bar{w}$).

The covariance term represents how fitness (w_i) varies with the value of the character (z_i) — if this term is positive, larger z values lead to higher fitness; if negative, smaller z values lead to higher fitness. The expectation term describes the fidelity or bias with which traits are transmitted to offspring. The terms can also be adapted to refer to groups containing altruists rather than individuals, in which case the covariance models the contribution of altruists to group fitness, while the expectation term models the loss due to the in-group loss of altruists. See **Group Selection**, **Kin Selection** and **Individual Selection**.

Production Rule A production rule is a condition-action (or if-then) rule that links an observation of the world (the condition) with an action to perform (the action). In agents, a set of production rules is typically used to link sensory data with motor function. See **Decision Tree**.

Prisoner's Dilemma The Prisoner's Dilemma is a game (in the game-theoretic sense) based on the hypothetical case of two prisoners, collaborators in some crime, who are separated and questioned by police. Each prisoner has two choices: to inform ("defect") or to stay silent ("cooperate"). If both stay silent, the prisoners receive a modest prison term; if both inform, they receive longer terms. However, if one informs while the other stays silent, the informant receives the minimum prison term while the other receives the maximum. The dominant strategy is to inform, since informing is rewarded with the shorter prison term regardless of what choice the other prisoner makes. However, iterating the game can lead to different conclusions. See **Cooperation** and **Stag Hunt**.

Pro-choice The position that a woman should have control over her own body during pregnancy and thus that she can choose to abort. See **Pro-life**.

Pro-life The position that human life begins at or just after conception and that abortion should be entirely prohibited or permitted only under extreme circumstances. See **Pro-choice**.

Punctuated Equilibrium Punctuated equilibrium refers to evolutionary histories in which long periods of morphological and behavioral stasis (or near stasis) are punctuated by short periods of rapid evolutionary change. Such punctuations often result in the appearance of new species. See **Phyletic Gradualism**.

Reciprocal Altruism Reciprocal altruism is the exchange of altruistic acts between individuals over time such that both individuals enjoy a net benefit. See **Cooperation** and **Altruism**.

Reduction A reduction of system A to system B is the provision of necessary and sufficient conditions for the properties and relations of A in terms of a different system B (the reduction base). In other words, there is an **isomorphism** between the two systems. (Cf. Batterman, 2007, on reductive bridge laws establishing synthetic type identities.) System B is typically taken as metaphysically more fundamental.

Examples include supposed reductions of biology to chemistry, of chemistry to physics, and of mental states to neurochemical states. See **Supervenience**.

Reflective Equilibrium Reflective equilibrium is a process of achieving a state of coherence between a core set of judgments about some domain, a theory about how those judgments should be made and wider theories relevant to the domain (e.g., including theories of human judgment).

Reinforcement Learning Reinforcement learning is a set of machine learning techniques for learning in a stochastic environment. An agent performs a sequence of actions affecting the environment. When the agent receives some (positive or negative) reward from the environment, it applies some algorithm to decide how much of the reward to attribute to the different actions leading to that algorithm (in a credit assignment process). The overall goal is to activate actions leading to positive rewards more often and suppress those leading to negative rewards.

Reproductive Strategy A reproductive strategy is the approach an organism takes to maximizing its genetic contribution to future generations by optimizing the division of parental investments amongst its expected offspring. When the environment is harsh and unpredictable, organisms will produce many offspring (since each will have little chance of surviving the environment) but invest little in each. When the environment is safe and predictable, organisms will predominantly compete with each other, and therefore invest a large (competitive) amount in each, and necessarily produce fewer offspring.

Sexual Dimorphism A species exhibits sexual dimorphism when its two sexes differ in morphology or behavior.

Sexual Selection Sexual selection refers to the competitive processes that occur within one sex for access to desirable members of the other sex. Typically, these processes are broken down into inter-male agression versus inter-female mate choice, but these sexual roles may often be reversed.

Simulation A (computer) simulation is a (computer) process that mimics features of a target physical process, such that a common dynamical theory is capable of describing both the simulation and its target process. For practical and theoretical reasons, a simulation is strictly simpler than its target, which entails a homomorphism from the target process to the simulation and the lack of an isomorphism. See **Homomorphism** and **Isomorphism**.

Social Simulation The field of social simulation employs simulations that contain societies of interacting agents, typically implemented in a bottom-up fashion, to explore social and societal phenomena. See **ALife**.

Sociobiology Sociobiology is a field that attempts to integrate a range of fields that study social behavior within both biology and sociology, including ethology, anthropology and behavioral economics. Sociobiology's approach to the study of social behavior is firmly rooted in evolution theory, and evolutionary psychology can be considered both a taxonomic and historical offshoot. See **Pop Sociobiology** and **Evolutionary Psychology**.

Species A species is commonly defined as a group of organisms that are capable of interbreeding and producing viable offspring. Especially for asexual species, alternative definitions are used in which similarity in genotype or phenotype is central.

Species Selection Selection which operates at the level of the species, in the form of extinctions and the propagation of new species. See **Multilevel Selection**.

Spontaneous Abortion Spontaneous abortion is the termination of a pregnancy that occurs via internal (non-intentional) causes. See **Induced Abortion**.

Stag Hunt The Stag Hunt is a game (in the game-theoretic sense) based on a scenario described in Rousseau's *The Social Contract*: two individuals on a hunt may choose either to hunt stag, which can only be successful if both cooperate, or to hunt hare, which can be successfully done alone. A stag yields more than twice the food of a hare, which implies that the game contains two Nash equilibria: either both cooperate to hunt stag (which yields a greater payoff) or both hunt hare (which involves less risk). See **Cooperation** and **Prisoner's Dilemma**.

Supervenience A system A is supervenient upon system B (the supervenience base) if and only if (a) B realizes (implements, instantiates) A; and (b) B is a member of a wider class of systems $\mathscr{B}$ any one of which could realize A. A is, therefore, said to be multiply realizable. In other words, there is a **homomorphism** from B to A.

(NB: Some people prefer to allow $\mathscr{B}$ to be a singleton set. This, however, loses the key characteristic of multiple realizability and conflates supervenience with **reduction**.)

Sympatric Speciation Speciation that occurs within a population in a single geographical area. Such speciation is considered rare, but may result if the population begins to exhibit polymorphic types that either cannot interbreed or find it difficult. See **Allopatric Speciation**.

Three Laws of Robotics The Three Laws of Robotics from Isaac Asimov's robot stories describe a set of "ethical" rules for robots to follow. With the addition of the later zeroeth law, the rules are as follows:

0. A robot may not harm humanity, or, by inaction, allow humanity to come to harm.
1. A robot may not injure a human being or, through inaction, allow a human being to come to harm.
2. A robot must obey any orders given to it by human beings.
3. A robot must protect its own existence.

Token An individual instance of some type of thing or event. Example: a dollar bill.

Type A collection of individual things or events characterizable by a set of properties held by all of those individuals. Example: a dollar.

Units of Selection See **Levels of Selection**.

Universal Computation, Universal Turing Machine (UTM) Devised by Alan Turing, a Turing Machine is a machine that can read and write symbols one at a time on an unbounded tape according to a table of rules. A Turing Machine is capable of performing any computation if given the right rules and symbol inputs. A Universal Turing Machine is capable of performing any computation (including the simulation of another Turing Machine) by changing the symbols on the tape alone.

GLOSSARY

Universalizable In ethics, a principle is universalizable if it can be adopted by everyone without logical inconsistency or absurd consequences. We appeal to universalizability when we ask, What if everyone did that?

Universal In evolutionary psychology, a behavior is universal if it is present in every member of the species (typically humans) or is pervasive within every group or culture, setting aside pathological cases.

Utilitarianism Utilitarianism is an ethical system which states that we should act so as to maximize the sum of expected utilities across a population. See **Utility** and **Consequentialism**.

Utility Utilities are theoretical entities used to explain agents' behavior under an assumption of Bayesian rationality — i.e., that agents aim to maximize their expected (probability-weighted) utilities. Utility functions map pairs of states of the world and actions to real numbers. A single unit of utility is sometimes called a "utile". Informally, utilities report the pleasantness or unpleasantness of the situation that an agent finds itself in, where "pleasant" is understood in a wide sense, incorporating any sensation that might have intrinsic value to the agent (e.g., the satisfaction of solving a problem would count as pleasant).

Utility Theory Utility theory is based on the principle that the set of preferences used in expected value calculations can be modeled by a cardinal (and potentially ratio-scale measurable) utility function. See **Utility** and **Decision Theory**.

Validation In simulation research validation refers to establishing (or testing) whether the simulation model corresponds to the targeted physical process. This corresponds (somewhat confusingly) to what Logical Positivists called **verification** and what in the philosophy of science generally is called confirmation. See **Verification** and **Confirmation**.

Verification In simulation research verification refers to establishing (or testing) whether the simulation model correctly implements the theory being investigated, including determining whether or not it is bug free. This usage is in contrast to that within the philosophy of science. See **Validation**.

Virtue Ethics Virtues are character traits that are considered either good in themselves or good due to their consequences. Virtue ethics suggests that goodness inheres in the character of a person.

Index

abortion, 91, 206
    action, 211
    and evolution, 206
    ethics of, 208, 228
    evolutionary stability of, 220
    examples in nature, 206
    simulation design, 210
action
    abortion, 91
    consensual mating, 189
    eating, 91
    movement, 91
    rape, 91, 186
    reproduction, 90
    resting, 91
    suicide, 91, 139
action rates, 95
actions, 90
adaptation hypothesis, 182
adoption queue, 108
agent
    age, 88
    behavior, 89
    genotypes, 91
    health, 88
    observations, 89
    utility, 88
Agent-Based Modelers (ABMers), 13
Agent-Based Models (ABMs), 82
agents, 88
aging
    adaptive theories of, 109
    comparison of hypotheses, 112
    experiments, 117
    non-adaptive theories of, 110
    simulation design, 112
    world, 112
allele, 9
altruism, 175
    biological, 8
altruistic suicide, 178
Amoeba, 80
antagonistic pleiotropy, 110, 115
approximation, 73
Aristotle, 4, 26
artificial intelligence (AI), 81
artificial life, 3, 12, 60, 77
Asimov, I., 32
Avida, 80
Axelrod, R., 83
Axtell, R., 84

Baldwin effect, 60, 81
Bentham, J., 27, 52
Bostrom, N., 45
bottom-up computer simulation (BUCS), 15
by-product hypothesis, 182
calibration, 74
cellular automata, 77
compatibility signature, 129
computation

limits of, 56, 76
Concorde fallacy, 149
confirmation, 71
consequentialism, 2, 27
Conway's Game of Life, 57, 77, 175
cooperation, 172
Cosmides, L., 34
crossover
    decision trees, 93
    production rules, 92
cultural evolution, 173
cycle, 87

Darwin, C.R., 1, 6, 7, 33, 41, 101, 145, 172
Dawkins, R., 9, 147, 149, 152
decision function, 89
decision tree, 93
defection, 173
demes, 100
demographics, 95
Dennett, D., 50
desertion hypothesis, 152
Diamond, J., 162, 169
discretization, 74
diversity hypothesis, 111

emergence, 61, 82
emergent property, 14, 60
environment of evolutionary adaptation (EEA), 15, 34, 35, 169, 183
epoch, 87
Epstein, J.M., 84
ethics, 3
    consequentialism, 27
    deontological, 26
    descriptive, 25
    evolutionary, 1, 41
    normative, 25

    of abortion, 208, 228
    simulating, 97
    virtue, 26
evolution, 6
    cultural, 173
    of aging, 108, 112, 125, 127, 134
    of altruism, 100, 103, 104, 106, 134, 175
    of parental investment, 145
    of suicide, 135
    of utility, 162
    simulated, 79
evolutionary ALife, 81
evolutionary ethics, 1
evolutionary psychology, 1, 81
    theories of rape in, 181
evolutionary stable strategy (ESS), 17, 135, 140
evolving psychology, 81
experiment
    as simulation, 71
experimental philosophy, 18

fitness, 7
    inclusive, 9, 100, 104
fitness function, 79
food, 87
food distribution function (fdf), 87
Franklin, A., 68
Frigg, R., 59, 73

Gap Theory of Utility, 164, 167
gene selection, 11
genetic algorithms (GAs), 79
genotypes, 91
Gilpin's predator-prey model, 101, 108
Grimm, V., 13, 66
group selection, 11, 100, 101, 137
groups, 100
    simulation design, 113

Hamilton, W.D., 37, 100, 104, 105
Hartmann, S., 5, 57, 59, 63
health, 88
hedonism, 28
hedonist rationality equation, 162
heritability, 6, 79
homomorphism, 64
host chromosome, 129
host vulnerability strings, 116
Hrdy, S.B., 208, 209
Hume, D., 4
Huxley, J., 41

inclusive fitness, 9, 100, 104
individual selection, 8
Individual-Based Modelers (IBMers), 13
Individual-Based Models (IBMs), 82
infection signature, 130
Iterated Prisoner's Dilemma (IPD), 16, 83, 172
    tournament, 83

James, W., 33, 47

Kant, I., 26, 47
kin selection, 9, 100, 104, 175
    "button", 108

Langton, C., 78
Lotka-Volterra equation, 82

Maynard Smith, J., 101
Medawar, P.B., 110
Mitteldorf's demographic theory, 111
modular mind, 34
Monte Carlo method, 58
Moore neighborhood, 89
mutation
    decision trees, 93
    meta-mutation, 94
    production rules, 92
mutation accumulation, 110, 116

Nash equilibrium, 17, 174
naturalistic fallacy, 1, 4, 37, 41

Ostrow, M., 42

Pandemonium, 78
parasite
    chromosome, 130
    transmission probability, 130
parental investment, 145, 189, 211
    simulation design, 148
Pascal's wager, 45
paternal uncertainty hypothesis, 155
Pavlov, 84
physical processes
    token, 63, 70
    type, 63, 70
Polyworld, 162
Popper, K., 5, 22, 43
positive association thesis, 163, 169
predator-prey
    Gilpin's model, 101, 108
predator-prey interactions, 111
predator-prey model, 177
Price equation, 105, 134
prior investment hypothesis, 149
production rule, 92, 189
punctuated equilibrium (PE), 103

Railsback, S., 66
Ramsey, F.P., 44
rape, 91, 181
    disutility of, 185
    simulation design, 186
    the unethical nature of, 184
    theories of in evolutionary psychology, 181
Ray, T., 80

Red Queen Hypothesis, 111
reductionism, 15
reflective equilibrium, 4, 19, 43, 48, 51, 53
Reiss, J., 59, 73
Repugnant Conclusion, 43
Ridley, M., 111

Samuelson, L., 164
selection, 79
    gene, 11
    group, 11, 100, 101, 137
    individual, 8
    kin, 9, 100, 104, 175
    levels of, 99
    species, 102, 127
self-age, 89
self-health, 89
self-sex, 90
Selfridge, O., 78
senescence, 99
sexually dimorphic behavior, 181
simulated evolution, 79
simulation, 11, 57
    ALife, 12, 60
    as experiment, 68
    computer, 55
    definition, 57, 63
    epistemology of, 70
    experimental, 20
    homomorphic, 64
Singer, P., 48, 52
Skulachev's phenoptosis theory, 111
Skyrms, B., 173
sociobiology, 37
speciation, 103, 131
species selection, 102, 127
    simulation design, 127
Stag Hunt, 173
statistics, 95

Sugarscape, 84
suicide, 84, 91, 135
    altruistic, 178
    simulation design, 137
    the evolutionary stability of, 140
Sumner, L.W., 209
supervene, 61
supervenience, 104
supervenient, 14
Swinkels, J., 164

Tierra, 80
time, 59
tit-for-tat (TFT), 17, 83, 173
token, 63
Tooby, J., 34
total utility, 95
Trivers, R.L., 37, 145, 149, 155, 159
type, 63

universality, 36
utilitarianism, 27, 29, 42
utility, 2, 5, 16, 17
    agent's, 88
    in agent-based modeling, 96
    total, 95

validation, 65, 70, 71
variation, 79
verification, 69, 70
virulence signature, 130
visualization, 73
von Neumann, J., 1, 13, 44, 58, 74, 77
vulnerability signature, 129

Wason's selection task, 35
Weismann hypothesis, 109, 112, 133
Weismann, A., 109
Williams, G.C., 101, 109
Wilson, D.S., 107

Wilson, E.O., 37, 42
Wynne-Edwards, V.C., 100, 104